ML

DO NOT REMOVE
CARDS FROM POCKET

Acknowledgments

I would like to thank the following manufacturers and organizations for supplying illustrations and information for this book: Brick Institute of America, American Plywood Association, Koppers Company, Teco Products and Testing Corporation, Stanley Tools, Skil Corporation, Hand Tools Institute, Crescent, Western Wood Products Association, Sears, Roebuck and Company and Jer Manufacturing, Inc. Their help is greatly appreciated.

Dedication

To Lydia, for allowing enough time to finish this book.

THE BACKYARD BUILDER'S BIBLE

BY CHARLES R. SELF

TAB TAB BOOKS Inc.
BLUE RIDGE SUMMIT, PA. 17214

FIRST EDITION

FIRST PRINTING—SEPTEMBER 1980

Library of Congress Cataloging in Publication Data

Self, Charles R.
  The backyard builder's bible.

  Includes index.
    1. Building—Amateurs' manuals. 2. Garden structures—Amateurs' manuals. I. Title.
  TH148.S43   690'.89   80-19809
  ISBN 0-8306-9702-0
  ISBN 0-8306-1285-8 (pbk.)

Cover photo courtesy of JER Manufacturing Inc.,
7205 Arthur Drive, Coopersville, Mich. 49404

Other TAB books by the author:

No. 872 *Wood Heating Handbook*
No. 892 *Do-It-Yourselfer's Guide To Chainsaw Use and Repair*
No. 949 *Do-It-Yourselfer's Guide To Auto Body Repair and Painting*
No. 1074 *How To Build Your Own Vacation Home*
No. 1144 *Working With Plywood, including Indoor/Outdoor Projects*
No. 1204 *The Brickworker's Bible*

2177264

Contents

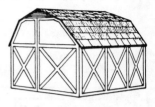

Introduction

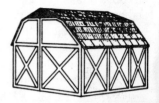

The reasons for adding outdoor structures to anyone's property will vary with needs, wallet and wants. You may need only a small shed to store a riding lawnmower and a few other lawn tools. A friend may need to house half a dozen horses or provide winter shelter for a couple of cows. Another neighbor or friend may wish a deck built on the back, side or front of his house so that summer parties are more comfortable and not confined to four walls. His best friend may prefer a patio for the same purpose.

The materials for doing these jobs range from simple wood framing through pole structures and on to brick and mortar. Naturally, the material selected will affect the tools needed, as well as the skills. In most construction work, care with materials, measurements and tools can help cut down on the need for the skill and art a pro spends years in acquiring, though the time needed to complete the work may be tripled or even quadrupled. A carefully done job will enhance property values and last for many, many years. A sloppily constructed outbuilding could well detract from the property value and, in extreme cases, the local building inspector could require you to tear down a poorly built structure.

Before starting any of the jobs covered in this book, I would recommend that you make a careful check of local building codes and talk with your building inspector. Too often homeowners, and sometimes even contractors, turn the relationship with the building inspector into an adverse one. That is neither necessary, nor a very good idea. In some areas, usually the most densely populated, many jobs can be done only by a licensed professional or under the supervision of one. Check before starting the job to find out what you're allowed to do and what codes you must follow. It would be nice to be able to list all the local codes for you, but the variance is so great and the changes so frequent that it is really impossible.

In the course of this book you'll find just about all you need to know to properly erect small barns, equipment sheds, decks, garages, storage sheds and gazebos. You will learn how to lay brick patios (the methods for which are easily adapted to slate and stone) and brick retaining walls. Everything from material to tool selection and use will be covered.

Some wiring, of the type generally designated as farm wiring, will also be examined. It is a great deal more pleasant to work in an outdoor building if you have a bit of light on the subject and a place to plug in that power drill or saw.

We'll also look at kits and semi-kits such as Jer barn kits, available in various sizes. The basic kit provides precut framing members, as well as all trim and shingles. The local lumber dealer supplies the needed plywood or other paneling to complete the job.

Special framing anchors such as those made by Teco can make the job of framing a lot easier for the amateur. These will be covered, as will other types of fasteners from common nails to screws and glues. We'll try to help you decide which tools are the best buys so that no more money is spent than is absolutely essential. A case in point is Stanley's new #19-035 miter box. Let's begin.

Charles R. Self

Chapter 1
Tools

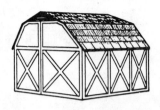

At the outset of the book, I might as well make a confession. I am a tool freak. If it were possible from a financial and storage standpoint, I would have at least one, and probably two or three, of every imaginable tool that might conceivably ever be of use. I try to stay out of hardware stores. Browsing through stores seriously endangers my financial health, and poring over tool catalogues has to be severely limited or I find myself with pen or phone in hand placing an order.

Fortunately, unlike some other addictions, this one offers some pleasant side benefits. I usually have just about any tool needed for a job at hand.

HAMMERS

Hammers are such simple tools. A hammer has a head with a driving or striking face on one end and a handle on the other. The correct hammer for the job being done can ease the work load considerably. It may seem silly for someone to have, as I do, claw hammers ranging from about 8 ounces in size up to 28 ounces. But a range of projects that I did, including a 12 by 24 foot deck, showed the advantage of the varied weights. In these cases the lighter hammers, those under 16 ounces, really weren't much use. When I was setting up and installing joist hangers (Teco's framing anchors) with their relatively short, special nails, the 16-ounce hammer was ideal. When I moved to the heavy framing timbers (4 × 6 pressure-treated pine, done with Kooper's outdoor treatment), the huge 28-ounce hammer was needed. Those 16 and 20 penny nails simply take too many blows with the lighter hammer, tiring your arm unnecessarily.

When the decking was installed, I moved back to Stanley's 22-ounce fiber glass handled hammer and used the same one for

Fig. 1-1. A carpenter's claw hammer.

installing the stairs. Railings were done with a True Temper 20 ounce hammer (Fig. 1-1). These were nailed with only 12 penny nails and then lag screwed to the deck facing. The trade off is head weight for number of blows needed to drive a nail. Since screw shanked nails were used exclusively for their greater holding power, I used a hammer a few ounces heavier than I would normally have. It is usually possible to sink a nail, with a 28-ounce hammer, in two blows. A 16-ounce one would do the hob in five or even six shots. The arithmetic is easy to figure. You lift and bring down a total of 56 ounces with the big hammer, while you lift and bring down a total of 80 or 96 ounces with the lighter hammer. That energy savings will really add up over the course of a day or two when you have to drive 35 pounds of nails!

Types of Hammers

Carpenter's hammers are often misused, most frequently as nail pullers, but just as often to strike surfaces for which they weren't designed. Partially driven nails may be pulled with a claw hammer if done properly (Figs. 1-2A and 1-2B). Larger nails may require a 3-foot crowbar for removal. Attempting to use a hammer to yank such nails is a waste of time. You're likely to do little more than snap the handle off. The striking surface of a carpenter's hammer is designed to drive nails. It can also be used to drive soft-handled chisels when cutting. Using the hammer to strike concrete or to drive cold chisels and masonry nails is a good way to ruin the face. Carpenter's hammers are designed to drive mild steel, not hardened steel. When other uses are contemplated, other hammers are needed. *Brick hammers, ball peens, engineer's hammers* and *sledge hammers* are designed to do a specific job. Your selection should be made accordingly.

Ball peen hammers are designed for striking cold chisels and punches, along with riveting and shaping metal. They can also be used for driving masonry nails.

Engineer's (blacksmith's) hammers come in head weights from 2 to 20 pounds and are used for moving heavy timbers. They can drive spikes, heavy cold chisels, and hardened nails. This type of sledge is *not* meant for striking stone or concrete.

Stone sledges and *spalling hammers* are designed for masonry work. The sledges break up stone. The spalling hammers are used for cutting and shaping (Fig. 1-3).

Bricklayer's hammers ahve a square, flat face with sharp corners and a blade with a sharpened edge. These hammers are specifically designed for setting and placing masonry units and cutting bricks.

Head Weights

Choosing a head weight for any job depends, first, on the length of the nails to be driven and, second, on your own strength and

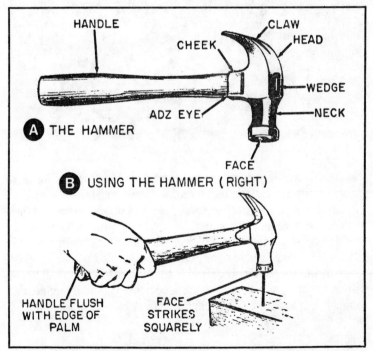

Fig. 1-2. The curved-claw nail hammer. (A). Hammer nomenclature (B). Proper use.

Fig. 1-3. Select the proper hammer for the job.

stamina. Generally, using the heaviest possible hammer for any particular job makes the most sense. However, trying to use a hammer too heavy to control simply messes up the surface of the work. If you have had little experience with carpenter's hammers, I would suggest starting with a 16-ounce hammer for general use and a 20 or 22-ounce hammer for heavier work, with possibly a 12-ounce model for trim work. Most of the time a 16-ounce ball been will do about all you'll need to do. Use a 2 or 3-pound engineer's hammer as a backup for the heavier stuff. A 16-ounce bricklayer's hammer is sufficient in most cases.

Handles

Handle material for hammers now includes solid steel, tubular steel, fiber glass and wood. The wood is generally hickory and is the easiest to replace in case of breakage. It is usually the most easily broken, but it also offers a good degree of shock absorption. Steel handles, whether solid or tubular, tend to be the hardest to snap off. They will bend, though but have the worst shock absorption characteristics, even with a rubber padding. Fiber glass is quite strong and absorbs shock about as well, possibly even a bit better than hickory. For general carpentry use, though, my preference is for fiber glass. I've never broken a fiber glass handle, and the material seems to tire me less during a long day of nail driving than does any other. If possible, borrow several hammers with different head weights and handle materials and try them for yourself. Then buy the one that suits you best (Fig. 1-4).

Fig. 1-4. Tools must be matched for the job.

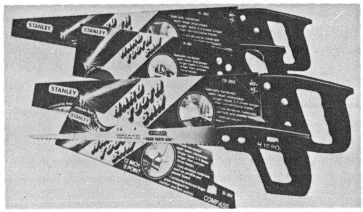

Fig. 1-5. This variety of saws is from a new series brought out by Stanley, using heat treatment on the teeth for longer life.

HANDSAWS

The variety of handsaws may seem a bit bewildering at first. We'll start with the crosscut saw, and go on from there.

Crosscut Saw and Ripsaw

The *crosscut saw*, as its name states, is used for cutting wood across the grain. Generally available with eight, 10 and 12 teeth per inch, such saws are essential ingredients of any toolbox where woodworking is contemplated. The handle will be of hardwood and should fit your hand comfortably. The number of teeth per inch you select has a direct bearing on the smoothness of the final cut. The more teeth per inch, the smoother the cut: the fewer teeth per inch, the quicker the cut. Select for quality, getting a tempered blade, taper ground, with bevel filed teeth (Fig. 1-5).

Ripsaws look exactly like crosscut saws at first glance, but a second glance will show some significant difference—fewer teeth per inch, for one—usually five or 5½. Teeth are sharpened straight across, so that they cut through wood as a chisel would. The ripsaw is designed for cutting with the grain of the wood.

Saw Care

A bit of information on the care of crosscut and ripsaws can be of help in preserving the life and holding the sharpness of the tools (Fig. 1-6). Because of the long blades of relatively thin steel, these saws, if not kept in a caprenter's toolbox with special slots, should be hung on the wall. They'll have holes in the top of the blade for that purpose. Laying them down allows the blades to warp and also puts them in a position where heavy objects can be set on them, warping the blades even more. Such a warped blade cannot help but provide you with a crooked cut. Use care when sawing in used wood where nails may be present.

Oil the saws after each use. I use a substance called *Tri-Flon* on my three best saws. Tri-Flon contains tiny particles of Teflon and does an exceptional job of protecting anything needing lubrication, from saws to pistols and rifles. Getting the gum and any possible dampness off the saws immediately cuts down on the chances of rusting. Remember, though, that only a very light film of oil is needed. Too much oil could get transferred to the wood being cut, causing difficulties in finishing the wood. Make sure you have sufficient height when cutting so the end of the saw won't contact the ground and bend the blade.

Backsaw, Compass Saw and Keyhole Saw

The *backsaw* is used with a miter box. In many respects it is similar to the crosscut saw, as it is designed to cut across the grain. In such use you will want accurate angles and, usually, a smooth finish cut. Most backsaws come with no fewer than 11 points to an inch; others have 13 or even 14.

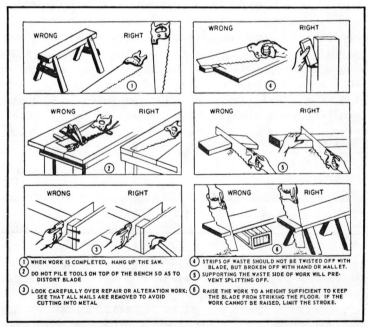

① WHEN WORK IS COMPLETED, HANG UP THE SAW.

② DO NOT PILE TOOLS ON TOP OF THE BENCH SO AS TO DISTORT BLADE

③ LOOK CAREFULLY OVER REPAIR OR ALTERATION WORK; SEE THAT ALL NAILS ARE REMOVED TO AVOID CUTTING INTO METAL

④ STRIPS OF WASTE SHOULD NOT BE TWISTED OFF WITH BLADE, BUT BROKEN OFF WITH HAND OR MALLET.

⑤ SUPPORTING THE WASTE SIDE OF WORK WILL PREVENT SPLITTING OFF.

⑥ RAISE THE WORK TO A HEIGHT SUFFICIENT TO KEEP THE BLADE FROM STRIKING THE FLOOR. IF THE WORK CANNOT BE RAISED, LIMIT THE STROKE.

Fig. 1-6. Take proper care of your saw.

Compass saws are designed for mild curved shape cuts in wood. Stanley's offering has a 12 inch blade and has either 12 or 14 teeth per inch. Nicholson offers several models with nine teeth per inch and in 12 or 14-inch blade lengths.

Keyhole saws are very similar in appearance to compass saws, but they feature a lighter blade with less depth so that tighter cuts can be made. Generally, a keyhole saw will have about 10 points to the inch and a 10-inch blade.

There is a little package known as the nest of saws that combines features you may well need. Looking at the Nicholson catalogue, I note that their offering comes with a 12-inch compass saw blade, nine teeth per inch; a 10-inch keyhole blade, 10 teeth per inch; and a metal blade, with 15 teeth per inch. Stanley's offering (or one of several, actually) has a 14-inch compass blade with eight points per inch, a 10-inch finishing blade with 10 points per inch, and a reversible blade, 16 inches long, with pruning teeth on one edge and an eight point crosscut on the reverse side.

Hacksaws

Hacksaws may not be essential for much of the work we'll be covering here. Sooner or later you'll find a need to cut a piece of pipe or plastic tubing, or slice through a nail. When that time comes, the hacksaw is nearly essential. Select the best frame you can find, one with a sturdy top and easy-to-handle blade attachments, along with a comfortable handle. Hacksaw blades come in a variety of lengths, so the frame should be adjustable to handle them. For rough work, you'll want a blade with possibly 14 teeth per inch; for finer work you'll find blades with 18, 24 and 32 teeth per inch.

SCREWDRIVERS

As Stanley's chart shows, there are five types of screwheads to be found in general use (Fig. 1-7). Actually, one is omitted, the Reed & Prince, but it is so seldom found these days it is worth ignoring. In fact, for almost all the types of work this book involves, you will need to consider only two of the types shown, the slotted and the Phillips heads. A good selection of screwdrivers is something everyone should have around the house no matter what work is expected to crop up, for there always seems to be something in need of tightening or loosening.

Most woodworking will be done with slotted screws, whether it is the installation of a metal patio door frame or some other project. Occasionally, you'll run into a need for Phillips head screws

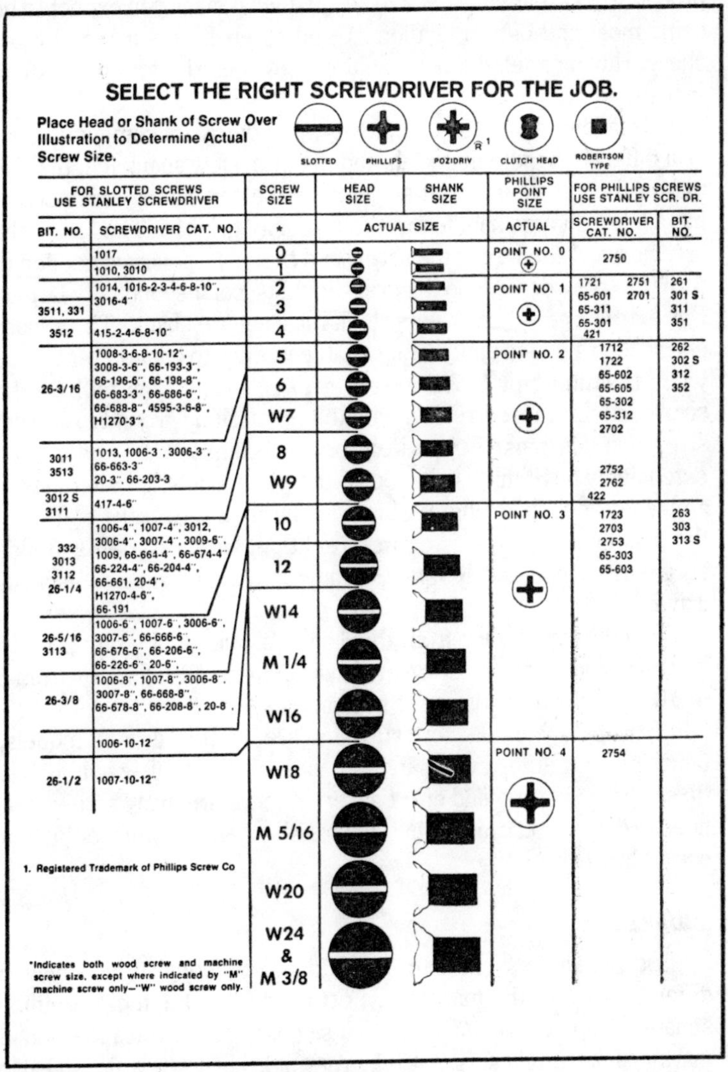

SELECT THE RIGHT SCREWDRIVER FOR THE JOB.

Place Head or Shank of Screw Over Illustration to Determine Actual Screw Size.

SLOTTED — PHILLIPS — POZIDRIV R[1] — CLUTCH HEAD — ROBERTSON TYPE

FOR SLOTTED SCREWS USE STANLEY SCREWDRIVER		SCREW SIZE	HEAD SIZE	SHANK SIZE	PHILLIPS POINT SIZE	FOR PHILLIPS SCREWS USE STANLEY SCR. DR.	
BIT. NO.	SCREWDRIVER CAT. NO.	*	ACTUAL SIZE		ACTUAL	SCREWDRIVER CAT. NO.	BIT. NO.
	1017	0			POINT NO. 0	2750	
	1010, 3010	1					
3511, 331	1014, 1015-2-3-4-6-8-10", 3016-4"	2			POINT NO. 1	1721 2751 65-601 2701 65-311	261 301 S 311
3512	415-2-4-6-8-10"	3 4				65-301 421	351
26-3/16	1008-3-6-8-10-12", 3008-3-6", 66-193-3", 66-196-6", 66-198-8", 66-683-3", 66-686-6", 66-688-8", 4595-3-6-8", H1270-3",	5 6 W7			POINT NO. 2	1712 1722 65-602 65-605 65-302 65-312 2702	262 302 S 312 352
3011 3513	1013, 1006-3", 3006-3", 66-663-3" 20-3", 66-203-3	8 W9				2752 2762	
3012 S 31¹¹	417-4-6"					422	
332 3013 3112 26-1/4	1006-4", 1007-4", 3012, 3006-4", 3007-4", 3009-6", 1009, 66-664-4", 66-674-4", 66-661, 20-4", H1270-4-6", 66-191	10 12			POINT NO. 3	1723 2703 2753 65-303 65-603	263 303 313 S
26-5/16 3113	1006-6", 1007-6", 3006-6", 3007-6", 66-666-6", 66-676-6", 66-206-6", 66-226-6", 20-6",	W14 M 1/4					
26-3/8	1006-8", 1007-8", 3006-8", 3007-8", 66-668-8", 66-678-8", 66-208-8", 20-8	W16					
	1006-10-12"				POINT NO. 4	2754	
26-1/2	1007-10-12"	W18					
		M 5/16					
		W20					
		W24 & M 3/8					

1. Registered Trademark of Phillips Screw Co

*Indicates both wood screw and machine screw size, except where indicated by "M" machine screw only—"W" wood screw only.

Fig. 1-7. Screwdriver types and sizes.

and their attendant screw drivers. Because there are only five sizes of Phillips head screwdrivers, it doesn't pay to just get a couple. Buy the five and you're ready. Slotted screws are another matter entirely. I'm not sure anyone knows just how many different slot sizes are produced, but as you can see the most used sizes range from the tiny 0 on up to the W24. Having a screwdriver for each size could mean at least 20 separate screwdrivers, so it is probably best that you check to see just what sizes you expect to be using most and buy to fit them. Good screwdrivers are no longer cheap. Having a selection you seldom if ever use is simply a waste of money.

To add to selection complexity, screwdrivers come with several different handle materials and in quite a few shank lengths. No matter the handle material selected, make certain the screwdriver has a large enough handle for it to be easily gripped. The larger the screwdriver, the larger the screw and the more power is needed to turn the screwdriver. So the grip should be correspondingly larger. For extra large screwdrivers, look for a square shank. The square shank allows you to apply an adjustable wrench to get greater torque when tightening or loosening screws. Basically, you will need a shank and tip of forged steel, with the tip machine ground to correct size. Handles must fit securely on the shanks. One style, used extensively by Stanley in its top-of-the-line models, is called *bolster* and really adds to strength. This is primarily a flare at the bottom of the handle that provides extra strength against twisting and side surges. Heat-treated *Boron steel* may be used in some top screwdrivers.

Lengths may vary from the 1-inch shank stubby on up to 16 inches or more for larger screwdrivers. Generally, lengths from 6 to 10 inches are handiest for normal use.

Stanley's Yankee screwdrivers are ratchet-driven models, taking several different bits in their chucks. While fairly expensive, these are great time and effort savers if there are many screws you need to drive. You can also buy square-shanked screwdriver bits for use in braces.

SQUARES

Squares are essential to any building job, though type may well differ because of the materials. For woodwork, the most common square used is the *combination*. This square offers you a method of marking for straight cuts or 45-degree angle cuts on boards, and also provides a 12-inch rule. Some squares have a bubble so that they may be used as a level. In my experience, such levels don't retain their accuracy very long because the combination square tends to get tossed around and battered a fair amount.

Try squares offer the 90-degree angle marking of the combination square and are more accurate, since there is no sliding part to wear. *Rafter* and *framing* squares are quite large and are needed for work on stairs and roof framing. It will take a bit of time to learn to use a rafter square, but the time spent is well worth it. Later, we'll look at using such a square to frame a gable roof.

LEVELS

Don't start any major job without a good level. By good, I mean a top quality model. Those levels with one vial and very light bodies just don't hold up well at all. They're not even worth having. Spend the few extra dollars and get the best you can afford, as the results will be well worth it.

A 2-foot square, such as my Stanley 100+ with an aluminum body, is essential for almost any work where plumbs or levels are needed. A 4-foot square, or even a 6-foot one, is a good idea when doors are being framed. The extra length provides greater accuracy over the longer distance (Fig. 1-8).

Mason's squares are usually of wood, normally mahogany, as is my 4-foot Montogomery Ward model. I also use this as a general 4-foot square and it makes a fine one. Still, the price is fairly high. If such a tool isn't used fairly frequently, it's best to save some money and stay with the good 2-foot level.

Line levels are needed when a line is run to a spot where a wall will be topping out, or to a point where several windows and doors will top out in a frame.

Fig. 1-8. A good level is just about essential to any kind of carpentry or masonry work.

PLANES AND SURFORMS

Planes are used to reduce the size of wood, smoothly, a little at a time. In most cases, the planes will be used on doors for the type of work we'll expect to do from this book. An adjustable *bench* or *jack plane* is the best bet. The jack plane will be about 14 inches long overall, while the bench plane won't quite reach 10 inches. Most planes are not used to cut end grain, though with care such cuts can be made.

Stanley's *Surform* tools for forming and cutting wood, plastic and light metals are among the handiest around. I've used mine for everything from rounding the edges on deck railings to cutting away excess body putty on auto repairs prior to final sanding. They're available in a wide variety of styles and shapes, and all have replaceable blades. In almost all woodworking, Surforms will replace files and can often be used instead of a plane for some jobs.

MEASURING DEVICES

The variety of *folding rules* and *tape measures* is another bewidlering array. The 6 or 8-foot folding rule is found in just about every carpenter's apron, and will be found in most mason's back pockets. Select one with a 6-inch brass extension for greatest utility.

Tape measures are another matter entirely. Selection isn't so simple because there are so many different length requirements. They are available from 6 feet to over 100 feet. One of the best moderate length tape measures is Stanley's Powerlock II. The inch-wide blade stays stiff for 7 feet, allowing you to almost dispense with the folding rule. The 1 inch blade Powerlock is 25 feet long, and narrower blades come in 16 and 20-foot lengths. You'll note I've said nothing about 6, 10 and 12-foot tapes. While not useless, the short length cuts them out for many jobs, and the few dollars saved probably isn't worth the lack of utility. Lufkin also makes fine measuring tools. I have one of their 100-foot tapes, which is handy when laying out fence lines.

Marking gauges feature a marked bar, with a sliding head and a pin in the head. For long cuts, you simply set the gauge and move it along the work, scoring as you go. Then cut along the line.

WRENCHES

In some assembly work, you'll need *wrenches* of one or two kinds even when working with wood. As an example, the ledger board of a deck will probably have to be fastened with lag bolts or screws to the sill of the house. A *ratchet* wrench can save a great deal of energy, though the job can be done as well but less rapidly with either a box or adjustable wrench. There is a wide variety of brand name wrenches. I would suggest wrenches made by the following manufacturers: Crescent, Bernzomatic, Craftsman, Montogomery Ward, Snap On and S-K. Be especially careful with cheap wrenches whose adjusters slip easily.

PLIERS

You may need several types of *pliers* when working on metal or wiring (Fig. 1-9). *Slip joint pliers* with two settings seem the most popular, for reasons that escape me. The groove joint slanted head pliers of the type made popular by Channelock are handier and provide greater holding power in virtually every instance. Again, stick with the known brand names.

Needle nose pliers are handy when doing any wiring job, for the tips are used to form the wire circles that fit on terminals. The

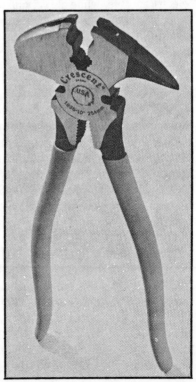

Fig. 1-9. These fence pliers serve many purposes from driving and pulling staples to cutting and stretching wire.

lightweight cutters can also be used for stripping wires (not cables, though, as they are not heavy enough for the job).

Side cutting pliers are needed for heavier wire and cable cutting chores. Like most pliers, they are available in several sizes. Generally an 8-inch pair will do most work around the home quite well.

Lineman's pliers are similar to side cutting pliers in their ability to cut heavier cable. They also offer a flat plier surface at the tip for bending jobs.

Locking pliers, commonly called vise-grips, are handy for holding items in place. Good quality is assured, again, by sticking to the known brand names.

KNIVES

Two types of knives are just about essential to construction projects. These are the *electrician's knife* and the *utility knife*.

The electrician's knife is a pocketknife with two blades. One blade is a more or less standard cutting blade. The other includes a screwdriver tip and a wire stripping groove and edge. In my experience, this is the best tool of all for stripping the outer sheathing from electrical cable, while also saving some fumbling for a screwdriver when it comes time to tighten terminals.

Utility knives have replaceable blades and serve many purposes. They can, in a pinch, strip cable, and are also useful for grooving wallboard to be snapped, among other things.

BRACES

While an electric drill saves effort and time, there are places where it can't be used because of a lack of power or other reasons. In such cases, the bit brace comes into its own. Generally available with 10 and 12-inch sweeps, a good bit brace will last just about forever. It provides drilling capacity in wood up to 1 inch. The auger bits should be of good quality, so you'll be sure the starter screw is accurately centered. High carbon steel is needed in the bits. Some hand finishing is a good idea for top quality, though you can probably get by without it. A recent check with Stanley Tools shows their top of the line 100+ bit set, with 13 auger bits in a wooden case, now retails for well over $100.

Electrican's bits are available for drilling in existing construction where long holes are needed. These are generally about 18 inches long and come in ¾-inch and 1-inch diameters. They can prove exceptionally handy when drilling through a girder or angling through a joist, into a floor and up through a sole plate. Other bits are far too short to complete the hole.

ELECTRIC DRILLS

One of the nicer things about power tools over the years has been the decrease in price. That's right, decrease. It may not seem so when you look at the prices of such drills now, with some models retailing for upwards of $65. The simple truth is that two decades ago a ⅜-inch chuck electric drill cost about $25-30 and offered variable speed but no reverse. Today, a similar drill lists for about $20, while the reversing model is about $30. Considering inflation over the years, that is indeed a price reduction. The $70 and $75 ⅜-inch drills today offer many more features, including rotary hammer action for easier drilling in masonry. Single speed drills with a ¼-inch capacity are under $10 (Fig. 1-10).

For light construction use, a middle range (power) ⅜-inch *chuck drill* should be powerful enough to do anything needed. Going to ½-inch chuck capacity gives a bit more versatility in heavy work but adds, usually, a fair amount to drill weight. My old ½-inch Craftsman weighs about 11 pounds, while a new ⅓ horsepower variable speed ⅜-inch model has a shipping weight of 4 pounds. Newer models than mine are lighter, for sure, but there is still a 5 pound or more weight difference.

Cordless drills are okay for light work where no more than 40-60 holes need to be drilled on a single charge. Otherwise, forget them. Sears does offer a ⅜-inch cordless drill that they claim will

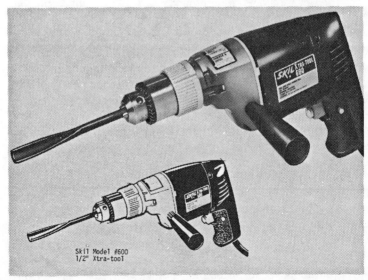

Skil Model #600
1/2" Xtra-tool

Fig. 1-10. Today's electric drill can be many tools in one.

provide 120 holes through a pine 2 × 4 on a single charge, but I have yet to use one. And they don't state hole size. Still, this particular model does come with the possibility of buying extra power packs. When buying any electrically driven tool, look for either double insulation or a grounding plug.

Twist drill bits for electric drills are available in both English and metric sizes. I would start with a small 10 or 12 bit set covering the most often used sizes, and progress bit by bit from there as the need arises. While the twist bits are not as expensive individually as auger bits for bit braces, the price can add up with a 115 bit set going from about $175 on up. Wood boring bits are also available, usually in a spade tip shape, but sometimes in an auger shape. There are also bits for masonry and nonferrous metals. Various accessories can be found for electric drills—buffers, sanders, right angle attachments, paint strippers and drill press stands. Select those as they are needed.

POWER SAWS

There are three primary types of electrically powered saws readily available to the homeowner. We'll start with the *circular saw*, since that is almost ubiquitous today.

Circular Saw

The saw should be either ground plug type or double insulated. New models have a button that must be pushed down before you can activate the on switch. The size of the saw blade is important, though less so in most cases than the power of the saw. The most popular size is 7 ¼ inches, which will do almost all modern framing chores. This size saw can also work on paneling and light plywood. In no case does it ever pay to buy a lightweight circular saw (except those specifically made to work with nothing but light plywood and paneling) for general use. The lightweight saw will heat up and lose power the first time you cut heavier framing members, causing binding in the cuts and all sorts of other problems.

For really heavy construction use, Skil makes three worm drive models that simply cannot be beaten for power over long periods of time (Fig. 1-11). Available in 6¼, 7¼ and 8¼-inch models, these worm drive saws retail for from about $165 to $190. Though expensive, they will prove well worth the money in the long run. If, though, all you're doing is putting up one small shed or deck, don't spend the money. Go instead for a Rockwell, Black & Decker, Craftsman or a Ward Power-Kraft 2 horsepower, 7¼-inch circular

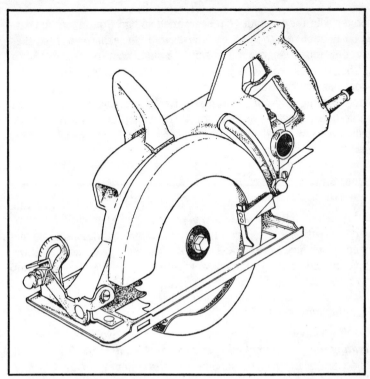

Fig. 1-11. A worm-driven circular saw.

saw and save almost $100. Anything under 2 horsepower generally
has too little power for working with material over the size of 2 ×
4s. The resultant binding can be hard on the wrist when the saw
kicks back out of a cut.

Circular saw blades come in a pretty fair variety of styles, each
meant for a particular use. Rip blades are used for the grain cuts.
Crosscut blades do cross grain cutting. Various types of combination blades do both rip and cross cutting. Plywood blades with
smaller teeth handle smoother cuts in finer materials. Cutoff
blades are used where there may be nails in the wood being cut.
They're a bit easier to resharpen than standard blades.

Carbide-Tipped Blades

A note on carbide-tipped blades is probably in order. I have
several, with the cost ranging from $18 to $22, against about $3.50
for standard circular saw blades (7¼-inch). They do stay sharp
longer than conventional blades, probably three to five times

longer. But they cannot be resharpened and give quite a rough cut. It costs about $1 to get a circular saw blade resharpened, so that three or four resharpenings bring the blade cost to about $6.50 to $7.50. In most cases, it pays to have five or six standard blades along and make the changes. Have the dull blades sharpened. I don't recall seeing any contractors supplying carbide-tipped blades to their men. Many people go for the 50 blade packages where individual blade cost is reduced, and each blade is simply discarded when dull.

Saber Saw

The *saber saw* is a handy tool for many jobs, though not particularly useful in the construction of outdoor buildings until you start working with shelving and other items where curved cuts may be needed. In fact, even the heaviest duty saber, or scroll, saws develop only about ½ horsepower. Most are strained to make cuts over an inch deep for any distance (Fig. 1-12).

Reciprocating Saw

For heavier work where your circular saw won't fit, the *reciprocating saw* is generally the way to go. With its 6 to 12-inch long blade, and ½ or more horsepower, it is excellent for many types of flush cuts and for getting in spots the circular saw simply cannot reach.

Radial Arm Saw

Stationary power tools can be exceptionally handy when repeat cuts are to be made. I've seen the time when I almost prayed for a radial arm saw while using a circular saw to make repetitive cuts. With the circular saw, each cut must be measured and marked. The *radial arm saw* allows you to set up guides to jigs and simply slap each board into place on the table and run the cut. Unfortunately, there's a trade off, as usual—price. A good 10-inch radial arm saw is going to run at least $300 and probably a fair bit more by the time you've gotten extra blades, a table and a few other accessories. It comes down to wasting time or spending money.

A 12-inch saw will run around $500 without a stand, so you can see that the expense really jumps when you start working with stationary tools. You can buy such saws for about $200, but I wouldn't really recommend it. The cut capacities are lower because of less power, and general quality is usually quite a bit lower.

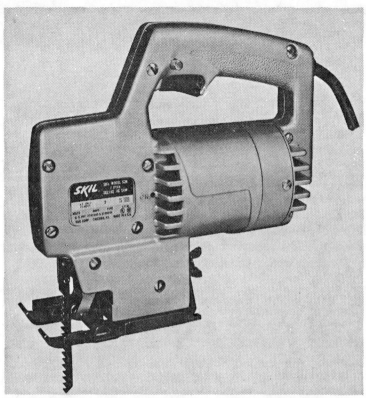

Fig. 1-12. A saber or jigsaw can be of help in cutting corners in light work.

If you decide to go for a radial arm saw, look for one with at least a 2 horsepower motor, a 3-inch cut capacity at 90 degrees, and the widest possible rip capacity (at least 24 inches if you wish to cut to the center of a 4 by 8-foot sheet of plywood at any time). A crosscut capacity of 15 inches or more is also handy as will be having all, or as many as possible, of the controls at or near the front of the saw.

Table saws offer many of the same features as the radial arm saw. If I had to choose just one, though, the radial arm saw would be it. There is the capacity to do end drilling and the ability to add router bits and shaper bits to form moldings. Plus, there is slightly greater versatility. The table saw, if properly set up, is a bit more accurate for really fine work, though.

MASONRY TOOLS

We've looked at many of the power and hand tools available for the home builder. I haven't bothered to cover masonry tools, since good quality trowels and pointing tools are readily available most everywhere. Other than that, you'll seldom need more than a contractor's wheelbarrow and a shovel and hoe. For large jobs, you may wish to rent a power mixer.

Chalk lines are invaluable for marking long cuts, or for marking shingle course runs. These may be the cased kind where the case holds the line and powdered chalk, or you can simply buy chunk chalk and pull a line through it.

Remington's power hammer can be handy when fastening wood or metal to concrete or cement block. This tool takes a .22 caliber blank and then shoots a hardened steel nail into the work surface when you tap the head with a hammer. The tool itself is not overly expensive ($30-$35), but the loads are too costly. In fact, .22 caliber long rifle bullets sell for about one-third or one-fourth the price of the blanks. The pins used are also costly, but they are made of special steel and are worth it. And the power hammer does a neater and stronger job than do masonry nails.

Chapter 2
Wood

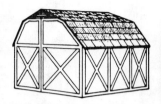

In general construction work, the selection of wood seems simple. Some sort of pine or fir is needed for framing, sheathing and finish. Plywood is also an alternative. There are various types of pine and fir and each accepts different loads, based on the inherent strength of the wood. Each type of wood needs some kind of preservative if exposed to the outdoors. Plywood comes in such a wide variety of styles and types it pays to know a bit about what you're buying. Moving to outdoor building, you'll also, in many areas, find you have an opportunity to purchase green lumber of various kinds for use on sheds and small buildings. It is possible to cure the green lumber, though many people go right ahead with it as it comes from the mill. Certain types should be avoided because of twisting problems, and some thought must always be given to shrinkage.

WOOD STRUCTURE

It pays to know a bit about the characteristics of wood in general and some peculiarities of particular species. Wood structure is shown in Fig. 2-1. The small cells in that illustration are arranged in rings, as in Fig. 2-2. *Sapwood* is the outer section of the tree, between the bark and the *heartwood*. Sapwood is lighter in color than heartwood, though the sapwood gradually changes to heartwood as the tree grows. Woods with small, closely packed cells are called *tight grained*, while woods with less tightly packed cells are called *open grained*.

Most wood today comes from the sawmill either slash or rift cut, with the majority of that probably being slash cut (Fig. 2-3). For the most attractive grain patterns, seldom a concern with outbuilding construction, one of the types of quartersawing is used (Fig. 2-4). Because quartersawing produces the most waste, it is seldom done for lumber being used in any but the finest furniture, and often

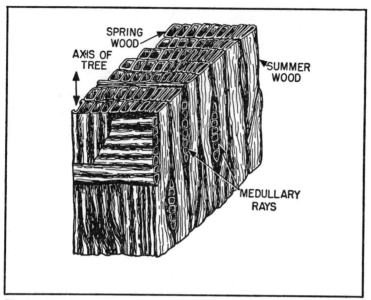

Fig. 2-1. Wood structure, showing medullary rays and spring and summer wood.

not even then. In fact, a quick check of local sawmills found only two owner/operators who claimed to know how to do quartersawing. Only one of those was willing to do some walnut for me that way.

Once cut to size, the lumber, if it is to be sold as standard construction lumber, is planed for smoothness, reducing the original full dimensions so that you now have nominally sized lumber.

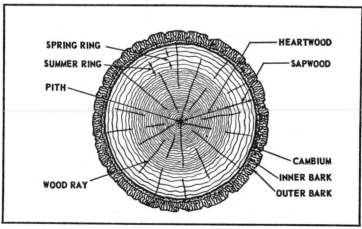

Fig. 2-2. Cross section of a tree.

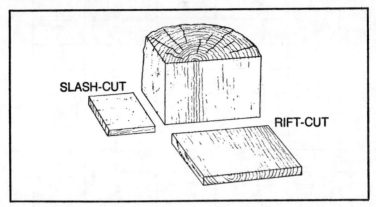

Fig. 2-3. Slash and rift cutting.

Drying reduces those dimensions a bit more. Not too long ago, nominal size requirements were reduced, so that a nominal 2 by 4 actually needs to measure only 1½ by 3½ inches. Essentially, you can reduce each dimension (except length) to arrive at the true lumber size when told the nominal size. A 4 by 12 would work out to 3½ by 11½ inches.

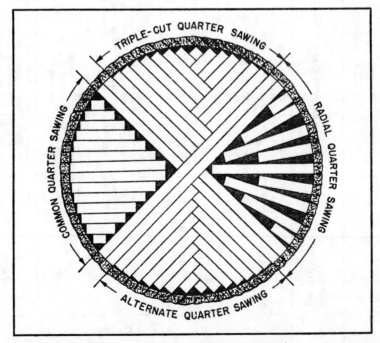

Fig. 2-4. Four different methods of quartersawing wood.

Table 2-1. Engineered Grades of Plywood.

	Grade Designation	Description and Most Common Use	Typical Grade-trademarks	Veneer Grade — Face	Veneer Grade — Back	Veneer Grade — Inner plies	Most Common Thicknesses (inch) (1)
Interior Type	C-D INT-APA	For wall and roof sheathing, subflooring, industrial uses such as pallets. Also available with intermediate glue or exterior glue. Specify intermediate glue for moderate construction delays; exterior glue for better durability in somewhat longer construction delays, and for treated wood foundations. (2) (10)	C-D 32/16 APA INTERIOR PS 1-74 000	C	D	D	5/16 3/8 1/2 5/8 3/4
	STRUCTURAL I C-D INT-APA and STRUCTURAL II C-D INT-APA	Unsanded structural grades where plywood strength properties are of maximum importance: structural diaphragms, box beams, gusset plates, stressed-skin panels, containers, pallet bins. Made only with exterior glue.	STRUCTURAL I C-D 24/0 APA INTERIOR PS 1-74 000 EXTERIOR GLUE	C(6) D(7)	D(7)	D(7)	5/16 3/8 1/2 5/8 3/4
	UNDERLAYMENT INT-APA	For underlayment or combination subfloor-underlayment under resilient floor coverings, carpeting in homes, apartments, mobile homes. Specify exterior glue where moisture may be present, such as bathrooms, utility rooms. Touch-sanded. Also available in tongue and groove. (2) (3) (9)	UNDERLAYMENT GROUP 2 APA INTERIOR PS 1-74 000	C plugged	D	C(8) & D	1/4 3/8 1/2 5/8 3/4
	C-D PLUGGED INT-APA	For built-ins, wall and ceiling tile backing, cable reels, walkways, separator boards. Not a substitute for UNDERLAYMENT as it lacks UNDERLAYMENT's indentation resistance. Touch-sanded. (2) (3) (9)	C-D PLUGGED GROUP 2 APA INTERIOR PS 1-74 000	C plugged	D	D	5/16 3/8 1/2 5/8 3/4
	2-4-1 INT-APA	Combination subfloor-underlayment. Quality base for resilient floor coverings, carpeting, wood strip flooring. Use 2-4-1 with exterior glue in areas subject to moisture. Unsanded or touch-sanded as specified. (2) (5) (11)	2-4-1 GROUP 1 APA INTERIOR PS 1-74 000	C plugged	D	C & D	1-1/8"

Grade	Description					1/4	5/16	3/8	1/2	5/8	3/4
C-C EXT-APA	Unsanded grade with waterproof bond for subflooring and roof decking, siding on service and farm buildings, crating, pallets, pallet bins, cable reels. (10)	C-C 42/20 EXTERIOR PS1*000 (APA)	C	C	C		5/16	3/8	1/2	5/8	3/4
STRUCTURAL I C-C EXT-APA and STRUCTURAL II C-C EXT-APA	For engineered applications in construction and industry where full Exterior type panels are required. Unsanded. See (9) for species group requirements.	STRUCTURAL I C-C 32/16 EXTERIOR PS1*000 (APA)	C		C		5/16	3/8	1/2	5/8	3/4
UNDERLAYMENT C-C Plugged EXT-APA	For underlayment or combination subfloor-underlayment under resilient floor coverings where severe moisture conditions may be present, as in balcony decks. Use for tile backing where severe moisture conditions exist. For refrigerated or controlled atmosphere rooms, pallets, fruit pallet bins, reusable cargo containers, tanks and boxcar and truck floors and linings. Touch-sanded. Also available in tongue and groove. (3) (9)	UNDERLAYMENT C-C PLUGGED GROUP 2 EXTERIOR PS1*000 (APA)	Plugged	C	$C^{(8)}$	1/4		3/8	1/2	5/8	3/4
C-C PLUGGED EXT-APA		C-C PLUGGED GROUP 3 EXTERIOR PS1*000 (APA)									
B-B PLYFORM CLASS I & CLASS II EXT-APA	Concrete form grades with high re-use factor. Sanded both sides. Mill-oiled unless otherwise specified. Special restrictions on species. Also available in HDO. (4)	B-B PLYFORM CLASS I EXTERIOR PS1*000 (APA)	B	B	C					5/8	3/4

(1) Panels are standard 4x8-foot size. Other sizes available.
(2) Also made with exterior or intermediate glue.
(3) Available in Group 1, 2, 3, 4, or 5.
(4) Also available in STRUCTURAL I.
(5) Made only in woods of certain species to conform to APA specifications.

(6) Special improved C grade for structural panels.
(7) Special improved D grade for structural panels.
(8) Special construction to resist indentation from concentrated loads.
(9) Also available in STRUCTURAL I (all plies limited to Group 1 species) and STRUCTURAL II (all plies limited to Group 1, 2, or 3 species).

(10) Made in many different species combinations. Specify by Identification Index.
(11) Can be special ordered in Exterior type for porches and patio decks, roof overhangs, and exterior balconies.

LUMBER GRADES

Grading of lumber takes into account several factors, including the number, size and spacing of knots, checks and splits. Most construction lumber today is kiln-dried—a process that will take, for softwoods, about three days. Kiln-dried lumber is generally specified to have from 12 to 15 percent moisture content.

Lumber grades for quality are of importance, and mill lumber (from mills that are members of specific associations with strict requirements to be met) will be grade-stamped. We won't look at the grade A to D lumbers for two reasons. They are nearly impossible to find these days and outrageously expensive when found; they are seldom used in construction, but more often in furniture making.

Number grading is of more importance to us. Number 1 Common is sound, tight-knotted lumber with few defects. It is hard to find.

Number 2 Common contains only a limited number of significant defects (larger knots, loose knots), but no major defects and no knotholes. It is easy to find but fairly expensive.

Number 3 Common has more defects than Number 2 Common and a few knotholes are allowed. Number 4 Common contains serious defects such as knotholes, checks and possibly even some decay.

Number 5 Common is not good lumber and should be used only where little or no structural strength is required. Also known as economy grade, it should only be used for such things as studding in a partition wall that bears no load at all.

For general use, Number 3 Common will suffice (also called standard grade by some associations). Number 2 Common has risen greatly in price lately.

PLYWOODS

Plywoods are essentially laminated woods in various thicknesses and overall sizes. They are easily available in 4 by 8, 4 by 12 and 4 by 16 sheets, depending on the particular style. Because of the variations in glue types and the physical properties of the wood used in each layer, the classifications of plywoods tend to get a bit more difficult than do those of standard lumbers.

Plywood offers a great many advantages in building. The layers are cross-plied for added strength. The large sheets can help in erecting a building quite rapidly. Plywood types which serve both as sheathing and exterior siding are also available.

Plywood starts its grading with interior and exterior types. Interior plywood requires adhesive durability of three types. If bonded with interior glue, it is not meant for exposure to wetting. Plywood bonded with intermediate glue will resist moderate high humidity and water leakage on a rather short term basis. The interior plywood may be bonded with exterior (waterproof) glue where there may be prolonged exposure to the elements. Exterior plywood must retain its glue bond under repeated wettings and dryings and is intended for use in a permanently exposed location.

There are also veneer grades, though we won't cover the fancier furniture veneer grades here. Grade A allows only neatly made repairs and has a smooth, paintable surface. Grade B provides a solid surface veneer, allowing only circular repairs and tight knots. Grade C allows knotholes to 1 inch, with limited splits, and is the minimimum permissible veneer grade for exterior types of plywood. Grade D permits large unplugged knotholes (up to 2½ inches) with limited splits.

Table 2-1 shows the applications for readily available interior and exterior grades of plywood. Table 2-2 shows the plywood grading for siding types meant to be used as finish siding.

A check of the charts will save you money in the long run, for there is simply no point in working with A-A exterior plywood when one side can't be seen or will be covered with hanging tools and shelves over much of its span. In fact, for most uses in outdoor buildings, A-C is about as good as you'll need to go. If you're working with a tool shed that will be partially or wholly screened by a fence, other structure or plants, then B-C will do the job. A couple of good coats of paint will cover any blemishes.

The same holds true for solid lumber, Tables 2-3 through 2-15 will help you estimate both the needed strength and the lowest grade of lumber possible to provide that strength, as well as the amount of lumber to buy. In every case, allow at least 10 percent for waste. Don'y overbuy to any great extent unless you're planning another project using similar materials in the near future.

TREATED LUMBER

There are many ways of treating lumber to resist rot and insect damage. Brush-on and soak-in solutions of many kinds are available. It is possible to select a type of lumber, usually cedar or redwood, with a natural resistance to decay. Selecting redwood or cedar may have drawbacks, although both woods are extremely attractive and easy to work with. Price can be the major drawback, while in some

Table 2-2. Appearance Grades of Plywood.

Grade Designation[2]	Description and Most Common Uses	Typical Grade-trademarks	Veneer Grade Face	Veneer Grade Back	Veneer Grade Inner Plies	Most Common Thicknesses (inch)[3]					
A-A EXT-APA	Use where appearance of both sides is important. Fences, built-ins, signs, boats, cabinets, commercial refrigerators, shipping containers, tote boxes, tanks, and ducts. (4)	A A G EXT APA PS 1 74	A	A	C	1/4		3/8	1/2	5/8	3/4
A-B EXT-APA	Use where the appearance of one side is less important. (4)	A B G EXT APA PS 1 74	A	B	C	1/4		3/8	1/2	5/8	3/4
A-C EXT-APA	Use where the appearance of only one side is important. Soffits, fences, structural uses, boxcar and truck lining, farm buildings, Tanks, trays, commercial refrigerators. (4)	A-C GROUP 1 EXTERIOR PS 1 74 000	A	C	C	1/4		3/8	1/2	5/8	3/4
B-B EXT-APA	Utility panel with solid faces. (4)	B B G EXT APA PS 1 74	B	B	C	1/4		3/8	1/2	5/8	3/4
B-C EXT-APA	Utility panel for farm service and work buildings, boxcar and truck lining, containers, tanks, agricultural equipment. Also as base for exterior coatings for walls, roofs. (4)	B-C GROUP 2 EXTERIOR PS 1 000	B	C	C	1/4		3/8	1/2	5/8	3/4

Type

Exterior

Type	Description	Grade stamp	A or B	A or B	C or C plgd	1/4	3/8	1/2	5/8	3/4
HDO EXT-APA	High Density Overlay plywood. Has a hard, semi-opaque resin-fiber overlay both faces. Abrasion resistant. For concrete forms, cabinets, counter tops, signs and tanks. (4)	HDO 60/60 BB PLYFORM EXT APA PS 1 74	A or B	A or B	C or C plgd		3/8	1/2	5/8	3/4
MDO EXT-APA	Medium Density Overlay with smooth, opaque, resin-fiber overlay one or both panel faces. Highly recommended for siding and other outdoor applications, built-ins, signs, and displays. Ideal base for paint. (4)	MDO BB G4 EXT APA PS 1 74	B	B or C	C		3/8	1/2	5/8	3/4
303 SIDING EXT-APA	Proprietary plywood products for exterior siding, fencing, etc. Special surface treatment such as V-groove, channel groove, striated, brushed, rough-sawn. (6)	303 SIDING 16 oc / GROUP 1 EXTERIOR PS 1 74 000 APA	(5)	C	C		3/8	1/2	5/8	
T 1-11 EXT-APA	Special 303 panel having grooves 1/4" deep, 3/8" wide, spaced 4" or 8" o.c. Other spacing optional. Edges shiplapped. Available unsanded, textured, and MDO. (6)	303 SIDING 16 oc / T1-11 / GROUP 1 EXTERIOR PS 1 74 000 APA	C or btr.	C	C				5/8	
PLYRON EXT-APA	Hardboard faces both sides, tempered, smooth or screened.	PLYRON EXT APA PS 1 74			C			1/2	5/8	3/4
MARINE EXT-APA	Ideal for boat hulls. Made only with Douglas fir or western larch. Special solid jointed core construction. Subject to special limitations on core gaps and number of face repairs. Also available with HDO or MDO faces.	MARINE AA EXT APA PS 1 74	A or B	A or B	B	1/4	3/8	1/2	5/8	3/4

(1) Sanded both sides except where decorative or other surfaces specified.
(2) Available in Group 1, 2, 3, 4, or 5 unless otherwise noted.
(3) Standard 4x8 panel sizes, other sizes available.
(4) Also available in Structural I (all plies limited to Group 1 species) and Structural II (all plies limited to Group 1, 2, or 3 species).
(5) C or better for 5 plies; C Plugged or better for 3-ply panels.
(6) Stud spacing is shown on grade stamp.
(7) For finishing recommendations, see form V307.
(8) For strength properties of appearance grades, refer to "Plywood Design Specification," form Y510.

Table 2-3. Thicknesses and Widths of Various Products.

	thickness in.	width in.		thickness in.	width in.
board lumber	1"	2" or more	beams & stringers	5" and thicker	more than 2" greater than thickness
light framing	2" to 4"	2" to 4"	posts & timbers	5" x 5" and larger	not more than 2" greater than thickness
studs	2" to 4"	2" to 4" 10' and shorter	decking	2" to 4"	4" to 12" wide
structural light framing	2" to 4"	2" to 4"	siding	thickness expressed by dimension of butt edge	
joists & planks	2" to 4"	6" and wider	mouldings	size at thickest and widest points	
Standard lengths of lumber generally are 6 feet and longer in multiples of 1'					

areas availability will be a problem. Because most cedar and all redwood is cut on the West Coast, easterners are going to have to pay more for the material. Transport costs are likely to keep rising steadily.

The major drawback of brush-on and soak-in solutions is that a small amount of the solution actually penetrates the wood to aid preservation. Pressure-treated wood of a local species is the answer. More costly than untreated wood, pressure-treated wood is

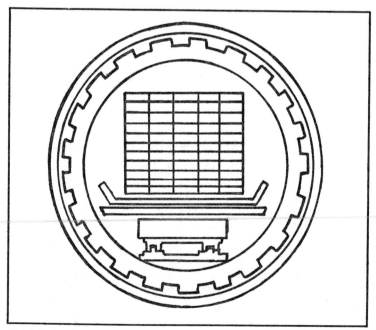

Fig. 2-5. Getting ready to pressure-treat wood.

Table 2-4. Standard Lumber Sizes Based on Western Wood Products Association Rules.

Product	Description	Nominal Size: Thickness In.	Nominal Size: Width In.	Dressed Dimensions: Thicknesses and Widths In. Surfaced Dry	Dressed Dimensions: Thicknesses and Widths In. Surfaced Unseasoned	Lengths Ft.
DIMENSION	S4S	2 3 4	2 3 4 6 8 10 12 Over 12	1-1/2 2-1/2 3-1/2 5-1/2 7-1/4 9-1/4 11-1/4 Off 3/4	1-9/16 2-9/16 3-9/16 5-5/8 7-1/2 9-1/2 11-1/2 Off 1/2	6 ft. and longer in multiples of 1′
SCAFFOLD PLANK	Rough Full Sawn or S4S	1¼ & Thicker	8 and Wider	Same	Same	6 ft. and longer in multiples of 1′
				Thickness In.	Width In.	
TIMBERS	Rough or S4S	5 and Larger		½ Off Nominal		6 ft. and longer in multiples of 1′

Product	Description	Nominal Size: Thickness In.	Nominal Size: Width In.	Dressed Dimensions: Thickness In.	Dressed Dimensions: Width In.	Lengths Ft.
DECKING Decking is usually surfaced to single T&G in 2″ thickness and double T&G in 3″ and 4″ thicknesses	2″ Single T&G	2	6 8 10 12	1½	5 6¾ 8¾ 10¾	6 ft. and longer in multiples of 1′
	3″ and 4″ Double T&G	3 4	6	2½ 3½	5¼	
FLOORING	(D & M), (S2S & CM)	3/8 1/2 5/8 1 1¼ 1½	2 3 4 5 6	5/16 7/16 9/16 3/4 1 1¼	1⅛ 2⅛ 3⅛ 4⅛ 5⅛	4 ft. and longer in multiples of 1′
CEILING AND PARTITION	(S2S & CM)	3/8 1/2 5/8 3/4	3 4 5 6	5/16 7/16 9/16 11/16	2⅛ 3⅛ 4⅛ 5⅛	4 ft. and longer in multiples of 1′
FACTORY AND SHOP LUMBER	S2S	1 (4/4) 1¼ (5/4) 1½ (6/4) 1¾ (7/4) 2 (8/4) 2½ (10/4) 3 (12/4) 4 (16/4)	5 and wider (4″ and wider in 4/4 No. 1 Shop and 4/4 No. 2 Shop)	25/32 (4/4) 1 5/32 (5/4) 1 13/32 (6/4) 1 19/32 (7/4) 1 13/16 (8/4) 2⅜ (10/4) 2¾ (12/4) 3¾ (16/4)	Usually sold random width	4 ft. and longer in multiples of 1′

Product	Description	Nominal Size: Thickness In.	Nominal Size: Width In.	Dressed Dimensions: Thickness In.	Dressed Dimensions: Width In.	Lengths Ft.
SELECTS AND COMMONS S-DRY	S1S, S2S, S4S, S1S1E, S1S2E	4/4 5/4 6/4 7/4 8/4 9/4 10/4 11/4 12/4 16/4	2 3 4 5 6 7 8 and wider	3/4 1 5/32 1 13/32 1 19/32 1 13/16 2 3/32 2⅜ 2 9/16 2¾ 3¾	1½ 2½ 3½ 4½ 5½ 6½ ¾ Off nominal	6 ft. and longer in multiples of 1′ except Douglas Fir and Larch Selects shall be 4′ and longer with 3% of 4′ and 5′ permitted.
FINISH AND BOARDS S-DRY	S1S, S2S, S4S, S1S1E, S1S2E These sizes also apply to alternate board grades.	3/8 1/2 5/8 3/4 1 1¼ 1½ 1¾ 2 2½ 3 3½ 4	2 3 4 5 6 7 8 and wider	5/16 7/16 9/16 5/8 3/4 1 1¼ 1⅜ 1½ 2 2½ 3 3½	1½ 2½ 3½ 4½ 5½ 6½ ¾ off nominal	3′ and longer. In Superior grade, 3% of 3′ and 4′ and 7% of 5′ and 6′ are permitted. In Prime grade, 20% of 3′ to 6′ is permitted.
RUSTIC AND DROP SIDING	(D & M) If ⅜″ or ½″ T & G specified, same over-all widths apply. (Shiplapped, ⅜-in. or ½-in. lap)	1	6 8 10 12	23/32	5⅜ 7⅛ 9⅛ 11⅛	4 ft. and longer in multiples of 1′
PANELING AND SIDING	T&G or Shiplap	1	6 8 10 12	23/32	5 7/16 7⅛ 9⅛ 11⅛	4 ft. and longer in multiples of 1′
CEILING AND PARTITION	T&G	5/8 1	4 6	5/16 23/32	3⅜ 5⅜	4 ft. and longer in multiples of 1′
BEVEL SIDING	Bevel or Bungalow Siding	½	4 5 6	15/32 butt, 3/16 tip	3½ 4½ 5½	3 ft. and longer in multiples of 1′
	Western Red Cedar Bevel Siding available in ½″, ⅝″, ¾″ nominal thickness. Corresponding thick edge is 15/32″, 9/16″ and ¾″. Widths for 8″ and wider, ½″ off nominal.	¾	8 10 12	¾ butt, 3/16 tip	7¼ 9¼ 11¼	3 ft. and longer in multiples of 1′

Product	Description	Nominal Size: Thickness In.	Nominal Size: Width In.	Thickness In. Surfaced Dry	Thickness In. Surfaced Green	Width In. Surfaced Dry	Width In. Surfaced Green	Lengths Ft.
STRESS RATED BOARDS	S1S, S2S, S4S, S1S1E, S1S2E	1 1¼ 1½	2 3 4 5 6 7 8 and Wider	¾ 1 1¼	25/32 1 1/32 1 9/32	1½ 2½ 3½ 4½ 5½ 6½ Off ¾	1 9/16 2 9/16 3 9/16 4⅝ 5⅝ 6⅝ Off ½	6 ft. and longer in multiples of 1′

ABBREVIATIONS
Abbreviated descriptions appearing in the size table are explained below.
S1S — Surfaced one side.
S2S — Surfaced two sides.
S4S — Surfaced four sides.
S1S1E — Surfaced one side, one edge.
S1S2E — Surfaced one side, two edges.
CM — Center matched.
D & M — Dressed and matched.
T & G — Tongue and grooved.
EV1S — Edge vee on one side.
S1E — Surfaced one edge.

See coverage estimator chart above for dressed Shiplap and Tongue and Groove (T&G) widths.

MINIMUM ROUGH SIZES Thicknesses and Widths Dry or Unseasoned All Lumber
80% of the pieces in a shipment shall be at least ⅛″ thicker than the standard surfaced size, the remaining 20% at least 3/32″ thicker than the surfaced size. Widths shall be at least ⅛″ wider than standard surfaced widths.
When specified to be full sawn, lumber may not be manufactured to a size less than the size specified.

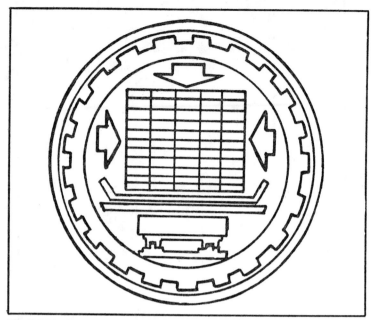

Fig. 2-6. Wolman preservative is forced into the wood cells.

usually around half or a bit less than half the cost of redwood or cedar. Pressure treating is nothing more than placing wood in pressurized cylinders and forcing chemicals into the wood (Figs. 2-5 and 2-6). The wood retains much more of the treatment chemicals and prevents the leaching of the chemicals into the soil. Koppers Company's *Outdoor* brand and *Wolman* brand pressure treatment chemicals have been in use for some time now and are very effective in preventing rot.

The treated wood is a rather odd green color on delivery, but will soon bleach to an attractive gray-silver color. Or it can be painted or stained just like untreated wood. The ease of workability is, of course, the same as for any other wood of the same species. The most popular woods for pressure treating are pines and firs, which are quite easily cut, nailed and otherwise shaped, formed or placed. Pressure-treated wood has been approved for use in home foundations, which gives some indication of just how well the treatment works (Fig. 2-7).

WOOD STRENGTH

For its weight, wood is one of the strongest materials generally available, with good resistance to stress. Because softwoods

LIGHT FRAMING and STUDS—2″ to 4″ Thick, 2″ to 4″ Wide
Recommended Design Values in Pounds Per Square Inch[1]

Species or Group	Grade	Extreme Fiber Stress in Bending "Fb"		Tension Parallel to Grain "Ft"	Horizontal Shear "Fv"	Compression		Modulus of Elasticity "E"
		Single	Repetitive			Perpendicular "Fc⊥"	Parallel to Grain "Fc"	Grades Described in Sections 40.00 and 41.00 WWPA 1970 Grading Rules
DOUGLAS FIR-LARCH	Construction[2]	1050	1200	625	95	385	1150	1,500,000
	Standard[2]	600	675	350	95	385	925	1,500,000
	Utility[2]	275	325	175	95	385	600	1,500,000
	Stud	800	925	475	95	385	600	1,500,000
DOUGLAS FIR SOUTH	Construction[2]	1000	1150	600	90	335	1000	1,100,000
	Standard[2]	550	650	325	90	335	850	1,100,000
	Utility[2]	275	300	150	90	335	550	1,100,000
	Stud	775	875	450	90	335	550	1,100,000
HEM-FIR	Construction[2]	825	975	475	75	245	925	1,200,000
	Standard[2]	475	550	275	75	245	725	1,200,000
	Utility[2]	225	250	125	75	245	500	1,200,000
	Stud	650	725	375	75	245	500	1,200,000
MOUNTAIN HEMLOCK	Construction[2]	875	1000	525	95	370	900	1,000,000
	Standard[2]	500	575	275	95	370	725	1,000,000
	Utility[2]	250	275	125	95	370	475	1,000,000
	Stud	675	775	400	95	370	475	1,000,000
MOUNTAIN HEMLOCK-HEM-FIR	Construction[2]	825	975	500	75	245	900	1,000,000
	Standard[2]	475	550	275	75	245	725	1,000,000
	Utility[2]	225	250	125	75	245	475	1,000,000
	Stud	650	725	375	75	245	475	1,000,000
WESTERN HEMLOCK	Construction[2]	925	1050	550	90	280	1050	1,300,000
	Standard[2]	525	600	300	90	280	850	1,300,000
	Utility[2]	250	275	150	90	280	550	1,300,000
	Stud	700	800	425	90	280	550	1,300,000
ENGELMANN SPRUCE-ALPINE FIR (Engelmann Spruce-Lodgepole Pine)	Construction[2]	700	800	400	70	195	675	1,000,000
	Standard[2]	400	450	225	70	195	550	1,000,000
	Utility[2]	175	200	100	70	195	375	1,000,000
	Stud	525	600	300	70	195	375	1,000,000
LODGEPOLE PINE	Construction[2]	775	875	450	70	250	800	1,000,000
	Standard[2]	425	500	250	70	250	675	1,000,000
	Utility[2]	200	225	125	70	250	425	1,000,000
	Stud	600	675	350	70	250	425	1,000,000
PONDEROSA PINE-SUGAR PINE (Ponderosa Pine-Lodgepole Pine)	Construction[2]	725	825	425	70	235	775	1,000,000
	Standard[2]	400	450	225	70	235	625	1,000,000
	Utility[2]	200	225	100	70	235	400	1,000,000
	Stud	550	625	325	70	235	400	1,000,000
IDAHO WHITE PINE	Construction[2]	675	775	400	70	190	725	1,200,000
	Standard[2]	375	425	225	70	190	650	1,200,000
	Utility[2]	175	200	100	70	190	425	1,200,000
	Stud	525	600	300	70	190	425	1,200,000
WESTERN CEDARS	Construction[2]	775	875	450	75	265	850	900,000
	Standard[2]	400	500	250	75	265	700	900,000
	Utility[2]	200	225	125	75	265	450	900,000
	Stud	600	675	350	75	265	450	900,000
WHITE WOODS (Western Woods)	Construction[2]	675	775	400	70	190	675	900,000
	Standard[2]	375	425	225	70	190	550	900,000
	Utility[2]	175	200	100	70	190	375	900,000
	Stud	525	600	300	70	190	375	900,000

[1] These design values apply to lumber when used at a maximum moisture content of 19% such as in most covered structures. For other conditions of use, see Section 31.00 of WWPA Grading Rules.
[2] Fb, Ft and Fc recommended design values apply only to 4″ widths of these grades.

Table 2-5. Design Values and Standard Grades for Light Framing and Studs.

Fig. 2-7. Select wood by grade marking.

(conifers) are most usually used for residential and outbuilding construction, the strength of these particular woods is of more interest than the strength of oaks, maples, elms, locusts and other hardwoods. There are several reasons for using softwood in building construction, starting with the greater ease of workability.

Not too long ago I helped a friend set up a long section of fence using seasoned locust posts and white oak boards. After about six hours of bending nails and having the oak boards split, we drove to town and purchased a Skil portable electric drill. There was simply no other way to prevent the oak from splitting. To make matters even worse, the only nails that would penetrate the locust posts with any ease were masonry nails. These nails provide little holding power and are shaped so as to make the splitting of the oak an even larger problem. Pines and firs don't create this problem, but they also don't provide the great final strength that the oak and locust combination does.

The simplest solution is to go one size larger in the posts and boards when using softwood so that strength is maintained. In general, hardwoods tend to cost more than do softwoods.

If green wood is to be used on any job, I would recommend that it be stacked and air-dried for a period of time before use. Select a flat site and stack the wood on cleats to keep it off the ground. Use

Table 2-6. Design Values and Standard Grades for Structural Light Framing.

STRUCTURAL LIGHT FRAMING and APPEARANCE—2" to 4" Thick, 2" to 4" Wide
Recommended Design Values in Pounds Per Square Inch[1]

Grades Described in Sections 42.00 and 50.00 WWPA 1970 Grading Rules

Species or Group	Grade	Extreme Fiber Stress in Bending "F_b" Single	Extreme Fiber Stress in Bending "F_b" Repetitive	Tension Parallel to Grain "F_t"	Horizontal Shear "F_v"	Compression Perpendicular "$F_{c\perp}$"	Compression Parallel to Grain "F_c"	Modulus of Elasticity "E"
DOUGLAS FIR-LARCH	Select Structural*	2100	2400	1200	95	385	1600	1,800,000
	No. 1*/Appearance	1750	2050	1050	95	385	1250/1500	1,800,000
	No. 2	1450	1650	1050	95	385	1050	1,600,000
	No. 3	800	925	475	95	385	600	1,500,000
DOUGLAS FIR SOUTH	Select Structural	2000	2300	1150	90	335	1400	1,400,000
	No. 1/Appearance	1700	1950	975	90	335	1150/1350	1,400,000
	No. 2	1400	1600	825	90	335	900	1,300,000
	No. 3	750	875	425	90	335	550	1,300,000
HEM-FIR	Select Structural	1650	1900	975	75	245	1300	1,500,000
	No. 1/Appearance	1400	1600	825	75	245	1050/1250	1,500,000
	No. 2	1150	1350	675	75	245	825	1,400,000
	No. 3	650	725	375	75	245	500	1,200,000
MOUNTAIN HEMLOCK	Select Structural	1750	2000	1000	95	370	1250	1,300,000
	No. 1/Appearance	1450	1650	900	95	370	1000/1200	1,300,000
	No. 2	1200	1400	700	95	370	775	1,100,000
	No. 3	675	775	400	95	370	475	1,000,000
MOUNTAIN HEMLOCK-HEM-FIR	Select Structural	1650	1900	975	75	245	1250	1,300,000
	No. 1/Appearance	1400	1600	825	75	245	1000/1200	1,300,000
	No. 2	1150	1350	675	75	245	775	1,100,000
	No. 3	650	725	375	75	245	475	1,000,000
WESTERN HEMLOCK	Select Structural	1800	2100	1050	90	280	1450	1,600,000
	No. 1/Appearance	1550	1800	900	90	280	1150/1350	1,600,000
	No. 2	1250	1450	750	90	280	900	1,400,000
	No. 3	700	800	425	90	280	550	1,300,000
ENGELMANN SPRUCE-ALPINE FIR (Engelmann Spruce-Lodgepole Pine)	Select Structural	1350	1550	800	70	195	950	1,300,000
	No. 1/Appearance	1150	1350	675	70	195	750/900	1,300,000
	No. 2	950	1100	550	70	195	600	1,100,000
	No. 3	525	600	300	70	195	375	1,000,000
LODGEPOLE PINE	Select Structural	1300	1500	875	70	250	1150	1,300,000
	No. 1/Appearance	1100	1300	725	70	250	900/1050	1,300,000
	No. 2	1050	1200	625	70	250	700	1,200,000
	No. 3	600	675	350	70	250	425	1,000,000
PONDEROSA PINE-SUGAR PINE (Ponderosa Pine-Lodgepole Pine)	Select Structural	1400	1650	825	70	235	1050	1,200,000
	No. 1/Appearance	1000	1400	700	70	235	850/1000	1,200,000
	No. 2	1000	1150	675	70	235	675	1,100,000
	No. 3	550	625	325	70	235	400	1,000,000
IDAHO WHITE PINE	Select Structural	1350	1550	775	70	190	1100	1,400,000
	No. 1/Appearance	1150	1300	650	70	190	875/1050	1,400,000
	No. 2	950	1050	625	70	190	675	1,300,000
	No. 3	525	600	300	70	190	425	1,200,000
WESTERN CEDARS	Select Structural	1500	1750	875	75	265	1200	1,100,000
	No. 1/Appearance	1300	1500	750	75	265	950/1100	1,100,000
	No. 2	1050	1200	625	75	265	750	1,000,000
	No. 3	600	675	350	75	265	450	900,000
WHITE WOODS (Western Woods)	Select Structural	1350	1550	875	70	190	950	1,100,000
	No. 1/Appearance	1150	1300	655	70	190	750/900	1,100,000
	No. 2	925	1050	550	70	190	600	1,000,000
	No. 3	525	600	300	70	190	375	900,000

[1]These design values apply to lumber when used at a maximum moisture content of 19% such as in most covered structures. For other conditions of use, see Section 140.00 of WWPA Grading Rules.

*For Dense values, see Table 8, page 10.

STRUCTURAL JOISTS and PLANKS and APPEARANCE—2″ to 4″ Thick, 6″ and Wider
Recommended Design Values in Pounds Per Square Inch[1]

Grades Described in Sections 62.00 and 50.00 WWPA 1970 Grading Rules

| Species or Group | Grade | Extreme Fiber Stress in Bending "Fb" | | Tension Parallel to Grain "Ft" | Horizontal Shear "Fv" | Compression | | | Modulus of Elasticity "E" |
		Single	Repetitive			Perpendicular "Fc⊥"	Parallel to Grain "Fc"		
DOUGLAS FIR-LARCH	Select Structural*	1800	2050	1200	95	385	1400		1,800,000
	No. 1/Appearance	1500	1750	1000	95	385	1250/1500		1,800,000
	No. 2*	1250	1450	825	95	385	1050		1,700,000
	No. 3	725	850	475	95	385	675		1,500,000
DOUGLAS FIR SOUTH	Select Structural	1700	1950	1150	90	335	1250		1,400,000
	No. 1/Appearance	1450	1650	975	90	335	1150/1350		1,400,000
	No. 2	1200	1350	775	90	335	950		1,300,000
	No. 3	700	800	450	90	335	600		1,100,000
HEM-FIR	Select Structural	1400	1650	950	75	245	1150		1,500,000
	No. 1/Appearance	1200	1400	800	75	245	1000/1250		1,500,000
	No. 2	1000	1150	650	75	245	875		1,400,000
	No. 3	575	675	375	75	245	550		1,200,000
MOUNTAIN HEMLOCK	Select Structural	1500	1700	1000	95	370	1100		1,300,000
	No. 1/Appearance	1250	1450	850	95	370	1000/1200		1,300,000
	No. 2	1050	1200	675	95	370	825		1,100,000
	No. 3	625	700	400	95	370	525		1,000,000
MOUNTAIN HEMLOCK-HEM-FIR	Select Structural	1400	1650	950	75	245	1100		1,300,000
	No. 1/Appearance	1200	1400	800	75	245	1000/1200		1,300,000
	No. 2	1000	1150	650	75	245	825		1,100,000
	No. 3	575	675	375	75	245	525		1,000,000
WESTERN HEMLOCK	Select Structural	1550	1800	1050	90	280	1300		1,600,000
	No. 1/Appearance	1350	1550	900	90	280	1150/1350		1,600,000
	No. 2	1100	1250	725	90	280	975		1,300,000
	No. 3	650	750	425	90	280	625		1,300,000
ENGELMANN SPRUCE-ALPINE FIR (Engelmann Spruce-Lodgepole Pine)	Select Structural	1200	1350	775	70	195	850		1,300,000
	No. 1/Appearance	1000	1150	675	70	195	750/900		1,300,000
	No. 2	850	950	525	70	195	625		1,100,000
	No. 3	475	550	300	70	195	400		1,000,000
LODGEPOLE PINE	Select Structural	1300	1500	875	70	250	1000		1,300,000
	No. 1/Appearance	1100	1300	750	70	250	900/1050		1,300,000
	No. 2	925	1050	600	70	250	750		1,200,000
	No. 3	525	625	350	70	250	475		1,000,000
PONDEROSA PINE-SUGAR PINE (Ponderosa Pine-Lodgepole Pine)	Select Structural	1200	1400	825	70	235	950		1,200,000
	No. 1/Appearance	1000	1200	700	70	235	850/1000		1,200,000
	No. 2	850	975	550	70	235	700		1,100,000
	No. 3	500	575	325	70	235	450		1,000,000
IDAHO WHITE PINE	Select Structural	1150	1300	775	70	190	950		1,400,000
	No. 1/Appearance	975	1100	650	70	190	875/1050		1,400,000
	No. 2	800	925	525	70	190	725		1,300,000
	No. 3	475	550	300	70	190	450		1,200,000
WESTERN CEDARS	Select Structural	1300	1500	875	75	265	1050		1,100,000
	No. 1/Appearance	1100	1300	750	75	265	950/1100		1,100,000
	No. 2	925	1050	600	75	265	800		1,000,000
	No. 3	525	625	350	75	265	500		900,000
WHITE WOODS (Western Woods)	Select Structural	1150	1300	775	70	190	850		1,100,000
	No. 1/Appearance	975	1100	650	70	190	750/900		1,100,000
	No. 2	800	925	550	70	190	625		1,000,000
	No. 3	475	550	300	70	190	400		900,000

[1]These design values apply to lumber when used at a maximum moisture content of 19% such as in most covered structures. For other conditions of use, see Section 140.00 of WWPA Grading Rules.
*For Dense values, see Table 8, page 10.

Table 2-7. Design Values and Standard Grades for Structural Joists and Planks.

Table 2-8. Design Values and Standard Grades for Beams and Stringers.

BEAMS and STRINGERS—5" and Thicker
Width More Than 2" Greater Than Thickness
Recommended Design Values in Pounds Per Square Inch[1]

Grades Described in Section 70.00 WWPA 1970 Grading Rules

Species or Group	Grade	Extreme Fiber Stress in Bending "Fb" (Single Members)	Tension Parallel to Grain "Ft"	Horizontal Shear "Fv"	Compression Perpendicular "Fc⊥"	Compression Parallel to Grain "Fc"	Modulus of Elasticity "E"
DOUGLAS FIR-LARCH	Select Structural* No. 1*	1600 1350	1050 900	85 85	385 385	1100 925	1,600,000 1,600,000
DOUGLAS FIR SOUTH	Select Structural No. 1	1550 1300	1050 850	85 85	335 335	1000 850	1,200,000 1,200,000
HEM-FIR	Select Structural No. 1	1250 1050	850 725	70 70	245 245	925 775	1,300,000 1,300,000
MOUNTAIN HEMLOCK	Select Structural No. 1	1350 1100	900 750	90 90	370 370	875 750	1,100,000 1,100,000
MOUNTAIN HEMLOCK—HEM-FIR	Select Structural No. 1	1250 1050	850 700	70 70	245 245	875 750	1,100,000 1,100,000
WESTERN HEMLOCK	Select Structural No. 1	1400 1150	950 775	85 85	280 280	1000 850	1,400,000 1,400,000
ENGELMANN SPRUCE-ALPINE FIR (Engelmann Spruce-Lodgepole Pine)	Select Structural No. 1	1050 875	700 600	65 65	195 195	675 550	1,100,000 1,100,000
LODGEPOLE PINE	Select Structural No. 1	1150 975	775 650	65 65	250 250	800 675	1,100,000 1,100,000
PONDEROSA PINE-SUGAR PINE (Ponderosa Pine-Lodgepole Pine)	Select Structural No. 1	1100 925	725 625	65 65	235 235	750 625	1,100,000 1,100,000
IDAHO WHITE PINE	Select Structural No. 1	1000 850	700 575	65 65	190 190	775 650	1,300,000 1,300,000
WESTERN CEDARS	Select Structural No. 1	1150 975	775 650	70 70	265 265	875 725	1,000,000 1,000,000
WHITE WOODS (Western Woods)	Select Structural No. 1	1000 850	700 575	65 65	190 190	675 550	1,000,000 1,000,000

[1] These design values apply to lumber when used at a maximum moisture content of 19% such as in most covered structures. For other conditions of use, see Section 140.00 of WWPA Grading Rules.
*For Dense values, see Table 8, page 10.

Table 2-9. Design Values and Grades for Decking.

decking

2″ to 4″ Thick. 4″ to 12″ Wide Design Values in Pounds Per Square Inch For Flatwise Use Only.

Species	Grade	DRY(1)		MC 15(2)	
		Extreme Fiber Stress in Bending "Fb" Repetitive	Modulus of Elasticity "E"	Extreme Fiber Stress in Bending "Fb" Repetitive	Modulus of Elasticity "E"
Douglas Fir—Larch	Selected Decking	2000	1,800,000	2150	1,900,000
	Commercial Decking	1650	1,700,000	1800	1,700,000
Douglas Fir-South	Selected Decking	1900	1,400,000	2050	1,500,000
	Commercial Decking	1600	1,300,000	1750	1,300,000
Hem-Fir	Selected Decking	1600	1,500,000	1700	1,600,000
	Commercial Decking	1350	1,400,000	1450	1,400,000
Mountain Hemlock	Selected Decking	1650	1,300,000	1800	1,300,000
	Commercial Decking	1400	1,100,000	1500	1,200,000
Mountain Hemlock—Hem-Fir	Selected Decking	1600	1,300,000	1750	1,300,000
	Commercial Decking	1350	1,100,000	1450	1,200,000
Western Hemlock	Selected Decking	1750	1,600,000	1900	1,700,000
	Commercial Decking	1450	1,400,000	1600	1,500,000
Engelmann Spruce, Alpine Fir (Engelmann Spruce—Lodgepole Pine)	Selected Decking	1300	1,300,000	1400	1,300,000
	Commercial Decking	1100	1,100,000	12⊂	1,200,000
Lodgepole Pine	Selected Decking	1450	1,300,000	1550	1,400,000
	Commercial Decking	1200	1,200,000	1300	1,200,000
Ponderosa Pine—Sugar Pine (Ponderosa Pine—Lodgepole Pine)	Selected Decking	1350	1,200,000	1450	1,300,000
	Commercial Decking	1150	1,100,000	1250	1,100,000
Idaho White Pine	Selected Decking	1300	1,400,000	1400	1.500,000
	Commercial Decking	1050	1,300,000	1150	1,400,000
Western Cedars	Selected Decking	1450	1,100,000	1550	1,100,000
	Commercial Decking	1200	1,000,000	1300	1,000,000
White Woods (Western Woods)	Selected Decking	1300	1,100,000	1400	1,100,000
	Commercial Decking	1050	1,000,000	1150	1,000,000

(1) DRY design values apply to lumber manufactured and used at a maximum moisture content of 19%. For other conditions of use, See Section 140.00, Para. 2, WWPA Grading Rules.

(2) MC 15 design values apply to lumber manufactured and used at a maximum moisture content of 15%. For other conditions of use, See Section 140.00, Para. 4, WWPA Grading Rules.

cleats between every layer of wood, spaced at 18 to 24-inch intervals, to allow air flow. Cover the top of the stack with plastic to keep rain or snow out, and let the wood sit for at least two months before use. Often people use green wood directly from the mill for outbuildings, and I've done the same. In such cases, it is imperative to realize that there will be both radial (width) and longitudinal shrinkage. The amount of shrinkage varies with wood type and can be considerable. In cases where green wood is used for siding, it is necessary to use the board and batten style of siding to allow for shrinkage in the width of the siding boards. No board over 6 inches wide should be used, as wider boards have a great tendency to cup and otherwise warp.

Oak is one of the worst of woods to use in its green state. It has a large tendency to twist as it dries, creating all kinds of problems. I have seen 2 by 6 oak boards pull 16 and 20 penny ring shanked nails a

Table 2-10. Design Values and Grades for Posts and Timbers.

POSTS and TIMBERS—5" x 5" and Larger
Width Not More than 2" Greater Than Thickness
Recommended Design Values in Pounds Per Square Inch[1]

Grades Described in Section 80.00 WWPA 1970 Grading Rules

Species or Group	Grade	Extreme Fiber Stress in Bending "Fb" Single Members	Tension Parallel to Grain "Ft"	Horizontal Shear "Fv"	Compression Perpendicular "Fc⊥"	Compression Parallel to Grain "Fc"	Modulus of Elasticity "E"
DOUGLAS FIR-LARCH	Select Structural* / No. 1*	1500 / 1200	1000 / 825	85 / 85	385 / 385	1150 / 1000	1,600,000 / 1,600,000
DOUGLAS FIR SOUTH	Select Structural / No. 1	1400 / 1150	950 / 775	85 / 85	335 / 335	1050 / 925	1,200,000 / 1,200,000
HEM-FIR	Select Structural / No. 1	1200 / 950	800 / 650	70 / 70	245 / 245	975 / 850	1,300,000 / 1,300,000
MOUNTAIN HEMLOCK	Select Structural / No. 1	1250 / 1000	825 / 675	90 / 90	370 / 370	925 / 800	1,100,000 / 1,100,000
MOUNTAIN HEMLOCK—HEM-FIR	Select Structural / No. 1	1200 / 950	800 / 650	70 / 70	245 / 245	925 / 800	1,100,000 / 1,100,000
WESTERN HEMLOCK	Select Structural / No. 1	1300 / 1050	875 / 700	85 / 85	280 / 280	1100 / 950	1,400,000 / 1,400,000
ENGELMANN SPRUCE-ALPINE FIR (Engelmann Spruce-Lodgepole Pine)	Select Structural / No. 1	975 / 800	650 / 525	65 / 65	195 / 195	700 / 625	1,100,000 / 1,100,000
LODGEPOLE PINE	Select Structural / No. 1	1100 / 875	725 / 600	65 / 65	250 / 250	850 / 725	1,100,000 / 1,100,000
PONDEROSA PINE—SUGAR PINE (Ponderosa Pine-Lodgepole Pine)	Select Structural / No. 1	1000 / 825	675 / 550	65 / 65	235 / 235	800 / 700	1,100,000 / 1,100,000
IDAHO WHITE PINE	Select Structural / No. 1	950 / 775	650 / 525	65 / 65	190 / 190	800 / 700	1,300,000 / 1,300,000
WESTERN CEDARS	Select Structural / No. 1	1100 / 875	725 / 600	70 / 70	265 / 265	925 / 800	1,000,000 / 1,000,000
WHITE WOODS (Western Woods)	Select Structural / No. 1	950 / 775	650 / 525	65 / 65	190 / 190	700 / 625	1,000,000 / 1,000,000

[1] These design values apply to lumber when used at a maximum moisture content of 19% such as in most covered structures. For other conditions of use, see Section 140.00 of WWPA Grading Rules.
*For Dense values, see Table 8, page 10.

47

LIGHT FRAMING—2″ and Less in Thickness, 2″ Wide
Recommended Design Values in Pounds Per Square Inch

Species or Group	Grade
DOUGLAS FIR-LARCH	Construction Standard Utility
DOUGLAS FIR SOUTH	Construction Standard Utility
HEM-FIR	Construction Standard Utility
MOUNTAIN HEMLOCK	Construction Standard Utility
MOUNTAIN HEMLOCK- HEM-FIR	Construction Standard Utility
WESTERN HEMLOCK	Construction Standard Utility
ENGELMANN SPRUCE— ALPINE FIR (Engelmann Spruce- Lodgepole Pine)	Construction Standard Utility
LODGEPOLE PINE	Construction Standard Utility
PONDEROSA PINE-SUGAR PINE (Ponderosa Pine Lodgepole Pine)	Construction Standard Utility
IDAHO WHITE PINE	Construction Standard Utility
WESTERN CEDARS	Construction Standard Utility
WHITE WOODS (Western Woods)	Construction Standard Utility

Grades Described
in Section 40.00

Extreme Fiber Stress in Bending "Fb"		Tension Parallel to Grain "Ft"	Compression Parallel to Grain "Fc"
Single	Repetitive		
950	1100	500	1150
450	500	225	925
125	150	75	375
900	1050	475	1000
425	475	225	850
125	150	75	350
750	875	400	925
350	400	175	775
100	125	50	325
800	925	425	900
375	425	200	725
125	125	50	300
750	875	400	900
350	400	175	725
100	125	50	300
825	950	450	1050
375	450	200	850
125	125	75	350
625	725	325	675
300	325	150	550
100	100	50	225
700	800	375	800
325	375	175	675
100	125	50	275
650	750	350	775
300	350	150	625
100	100	50	250
600	700	350	775
275	325	175	650
75	100	50	275
700	800	375	850
325	375	175	700
100	125	50	300
600	700	325	675
275	325	150	550
75	100	50	225

These design values apply to lumber when used at a maximum moisture content of 19%, such as in most covered structures. For other conditions of use, see Section 140.00 of WWPA Grading Rules.

Table 2-12. Design Values and Grades for Light Framing 3 Inches and Less in Thickness and 3 Inches Wide.

LIGHT FRAMING—3" and Less in Thickness, 3" Wide
Recommended Design Values in Pounds Per Square Inch[1]

Horizontal Shear "Fv", Compression Perpendicular "Fc ⊥" and Modulus of Elasticity "E" values are shown in Table 1, Light Framing.

Species or Group	Grade	Extreme Fiber Stress in Bending "Fb"		Tension Parallel to Grain "Ft"	Compression Parallel to Grain "Fc" Grades Described in Section 40.00
		Single	Repetitive		
DOUGLAS FIR-LARCH	Construction	875	1000	500	1150
	Standard	550	625	300	925
	Utility	150	175	100	450
DOUGLAS FIR SOUTH	Construction	825	950	475	1000
	Standard	525	600	300	850
	Utility	150	175	75	400
HEM-FIR	Construction	700	800	400	925
	Standard	425	500	250	775
	Utility	125	150	75	375
MOUNTAIN HEMLOCK	Construction	725	825	400	900
	Standard	450	525	250	725
	Utility	125	150	75	350
MOUNTAIN HEMLOCK-HEM-FIR	Construction	700	800	400	900
	Standard	425	500	250	725
	Utility	125	150	75	350
WESTERN HEMLOCK	Construction	750	875	425	1050
	Standard	475	550	275	850
	Utility	150	150	75	400
ENGELMANN SPRUCE-ALPINE FIR (Engelmann Spruce-Lodgepole Pine)	Construction	575	650	325	675
	Standard	350	425	200	550
	Utility	100	125	50	275
LODGEPOLE PINE	Construction	625	725	350	800
	Standard	400	450	225	675
	Utility	125	125	75	325
PONDEROSA PINE-SUGAR PINE (Ponderosa Pine-Lodgepole Pine)	Construction	600	675	325	775
	Standard	375	425	200	625
	Utility	100	125	50	300
IDAHO WHITE PINE	Construction	550	650	325	775
	Standard	350	400	200	650
	Utility	100	125	50	300
WESTERN CEDARS	Construction	625	725	350	850
	Standard	400	450	225	700
	Utility	125	125	75	325
WHITE WOODS (Western Woods)	Construction	550	650	325	675
	Standard	350	400	200	550
	Utility	100	125	50	275

[1] These design values apply to lumber when used at a maximum moisture content of 19%, such as in most covered structures. For other conditions of use, see Section 140.00 of WWPA Grading Rules.

Table 2-13. Design Values for Dense Douglas Fir.

DENSE DOUGLAS FIR—LARCH
Recommended Design Values in Pounds Per Square Inch[1]

Grades Described in Section 53.00

Species or Group	Grade	Extreme Fiber Stress in Bending "Fb" Single	Extreme Fiber Stress in Bending "Fb" Repetitive	Tension Parallel to Grain "Ft"	Horizontal Shear "Fv"	Compression Perpendicular "Fc⊥"	Compression Parallel to Grain "Fc"	Modulus of Elasticity "E"
STRUCTURAL LIGHT FRAMING	Dense Sel. Struc.	2450	2800	1400	95	455	1850	1,900,000
	Dense No. 1	2050	2400	1200	95	455	1450	1,900,000
	Dense No. 2	1700	1950	1000	95	455	1150	1,700,000
STRUCTURAL JOISTS AND PLANKS	Dense Sel. Struc.	2100	2400	1400	95	455	1650	1,900,000
	Dense No. 1	1800	2050	1200	95	455	1450	1,900,000
	Dense No. 2	1450	1700	950	95	455	1250	1,700,000
BEAMS AND STRINGERS	Dense Sel. Struc.	1900		1250	85	455	1300	1,700,000
	Dense No. 1	1550		1050	85	455	1100	1,700,000
POSTS AND TIMBERS	Dense Sel. Struc.	1750		1150	85	455	1350	1,700,000
	Dense No. 1	1400		950	85	455	1200	1,700,000

These design values apply to lumber when used at a maximum moisture content of 19%, such as in most covered structures. For other conditions of use, see Section 140.00 of WWPA Grading Rules.

Table 2-14. Grade Selector Charts.

			SPECIFICATION CHECK LIST
APPEARANCE GRADES	**SELECTS**	B & BETTER (IWP—SUPREME) C SELECT (IWP—CHOICE) D SELECT (IWP—QUALITY)	☐ Grades listed in order of quality. ☐ Include all species suited to project. ☐ Specify lowest grade that will satisfy job requirement. ☐ Specify surface texture desired. ☐ Specify moisture content suited to project. ☐ Specify ⓌⓅ grade stamp. For finish and exposed pieces, specify stamp on back or ends.
	FINISH	SUPERIOR PRIME E	
	PANELING	CLEAR (ANY SELECT OR FINISH GRADE) NO. 2 COMMON SELECTED FOR KNOTTY PANELING NO. 3 COMMON SELECTED FOR KNOTTY PANELING	
	SIDING (BEVEL, BUNGALOW)	SUPERIOR PRIME	

	**BOARDS
	SHEATHING
	& FORM
	LUMBER**
		NO. 1 COMMON (IWP—COLONIAL)
		NO. 2 COMMON (IWP—STERLING)
		NO. 3 COMMON (IWP—STANDARD)
		NO. 4 COMMON (IWP—UTILITY)
		NO. 5 COMMON (IWP—INDUSTRIAL)
		
		ALTERNATE BOARD GRADES
		SELECT MERCHANTABLE
		CONSTRUCTION
		STANDARD
		UTILITY
		ECONOMY
			WESTERN RED CEDAR
			
			FINISH CLEAR HEART
			PANELING A
			AND CEILING B
			
			BEVEL CLEAR—V.G. HEART
			SIDING A — BEVEL SIDING
			         B — BEVEL SIDING
			         C — BEVEL SIDING

Table 2-15. Dimension Grades

LIGHT FRAMING 2" to 4" Thick 2" to 4" Wide	CONSTRUCTION STANDARD UTILITY ECONOMY*	This category for use where high strength values are NOT required; such as studs, plates, sills, cripples, blocking, etc.
STUDS 2" to 4" Thick 2" to 4" Wide	STUD ECONOMY STUD*	An optional all-purpose grade limited to 10 feet and shorter. Characteristics affecting strength and stiffness values are limited so that the "Stud" grade is suitable for all stud uses, including load bearing walls.
STRUCTURAL LIGHT FRAMING 2" to 4" Thick 2" to 4" Wide	SELECT STRUCTURAL NO. 1 NO. 2 NO. 3 ECONOMY*	These grades are designed to fit those engineering applications where higher bending strength ratios are needed in light framing sizes. Typical uses would be for trusses, concrete pier wall forms, etc.
APPEARANCE FRAMING 2" to 4" Thick 2" and Wider	APPEARANCE	This category for use where good appearance and high strength values are required. Intended primarily for exposed uses. Strength values are the same as those assigned to No. 1 Structural Light Framing and No. 1 Structural Joists and Planks.
STRUCTURAL JOISTS & PLANKS 2" to 4" Thick 6" and Wider	SELECT STRUCTURAL NO. 1 NO. 2 NO. 3 ECONOMY*	These grades are designed especially to fit in engineering applications for lumber six inches and wider, such as joists, rafters and general framing uses.

timbers

BEAMS & STRINGERS	SELECT STRUCTURAL NO. 1 NO. 2 (NO. 1 MINING) NO. 3 (NO. 2 MINING)	POSTS & TIMBERS	SELECT STRUCTURAL NO. 1 NO. 2 (NO. 1 MINING) NO. 3 (NO. 2 MINING)

*Design values are assigned to all dimension grades except Economy.

good inch out of the studs to which they were nailed after only a couple of months in place.

Softwood strength varies a good deal, though in most cases substitutions are relatively easily made. Remember that the weakest species are the hemlocks, Engelmann spruce, subalpine firs (weakest of all) and Ponderosa pine. Strongest of all are Douglas fir-larch and Southern pine. As an example, a 2 by 6 of Southern pine can be used as a floor joist over a span of 8 feet 10 inches. Engelmann spruce can only span 7 feet 7 inches, and subalpine firs span a meager 7 feet. The span will meet the generally required deflection needs of the span in inches divided by 360, with a 10 pound dead load and a 40 pound live load, using ¾-inch plywood nailed to the joists.

Such strength requirements may not seem critical to you. But I have no more desire to have a tool shed or barn floor collapse under me than I do the floor of my home. In cases where workshop use is intended and heavy tools may be added, it might well pay you to raise the size of your floor joists one or two sizes to add strength to the floor.

OTHER MATERIALS

While wood is the primary framing and general building, material for outbuildings, you'll soon run into a variety of others like zinc-coated studs, rafters and joists, fiber glass roof panels, brick and stone. Each of these materials will be covered with reasonable thoroughness in separate chapters.

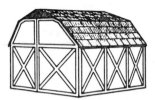

Chapter 3
Fastening &
Joining Wood

The art of joining wood usually requires one of three items: a nail a screw, glue or some combination of the three. Often in finish carpentry for shelving and cabinets, specially shaped joints are used to add strength to a simple glued joint that will be placed under stress. In rough carpentry (framing work), the most frequent joint design is the simple butt joint where one board meets another and is nailed in place. The meeting is usually in the shape of a "T" as when studs are nailed to sole and top plates. Lap joints are often used when joists pass over a girder in a basement, and a bird's mouth cutout may be utilized when rafters pass over wall top plates to form an eave overhang for a roof.

NAILS

Most nails used in construction today are wire nails of mild steel. They come in a wide variety of types and sizes (Fig. 3-1 and Table 3-1). A few general rules in selecting nails may be of some assistance, when applied to the size chart in Table 3-2. Nails must be three times the length of the material they are intended to hold, so that two-thirds of the length of the nail is driven into the second piece of material to provide secure anchorage. Best security is provided if the nails are driven at a slight angle towards each other. Proper placement helps, too—nails driven with the grain of the wood do not hold as well as those driven across the grain.

Types of Nails

Common wire nails are used for building framing and general rough carpentry including, in many cases, the installation of certain types of wood siding.

Finishing nails are made of finer wire than are similar sized common wire nails. The nails have very small heads so they can be easily sunk below the surface of the work. Finishing nails are used

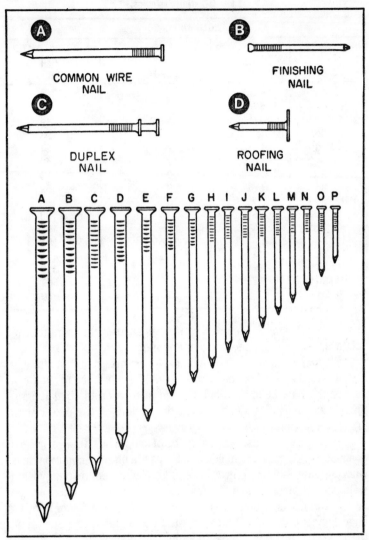

Fig. 3-1. Common wire types and their sizes.

most often for exterior and interior finish work and can be found in color matches for different kinds of wood wall paneling.

Duplex nails are double headed and have two primary uses. They are handy for nailing temporary bracing as the lower head holds the brace tight, while the upper head makes the nail easy to pull out. Too, duplex nails are used to hold electric fence insulators, either plastic or ceramic. If you use an electric fence, go with plastic

Table 3-1. Nail Statistics.

	LENGTH AND GAGE		
SIZE	INCHES	NUMBER	APPROXIMATE NUMBER TO POUND
A 60d	6	2	11
B 50d	5½	3	14
C 40d	5	4	18
D 30d	4½	5	24
E 20d	4	6	31
F 16d	3½	7	49
G 12d	3¼	8	63
H 10d	3	9	69
I 9d	2¾	10¼	96
J 8d	2½	10¼	106
K 7d	2¼	11½	161
L 6d	2	11½	181
M 5d	1¾	12½	271
N 4d	1½	12½	316
O 3d	1¼	14	568
P 2d	1	15	876

insulators. A single hammer shot will shatter the ceramic insulators, and we all miss once in a while.

Roofing nails have diamond-shaped points, round shafts and are either galvanized or coated with a tar-like substance to cut down on rusting. The heads are fairly large in relation to shaft length. Shingle nailing is always begun at the middle of the material to prevent buckling.

In addition to these general types of nails, you'll find variations of annular and screw shanked nails for use in place of common nails where greater holding power is needed. Nails used to hold gypsum wallboard in place should be annular shanked so as to cut down on nail popping. It is a good idea to use screw shanked nails for flooring underlayment and sometimes for the flooring itself (depending on the type) to cut down on nail popping and squeaking.

In addition to steel nails, aluminum nails are now available, though in some areas they're a bit hard to locate. Aluminum nails, most especially in exterior construction, offer some advantages over steel nails since they can't rust and mar the finish of a fence, deck or other structure. Even galvanized nails will eventually allow rust streaks to penetrate. You knock most of the zinc coating off the head of galvanzied nails as you drive them, and the heads get the greatest amount of exposure to the elements. Even when available,

Table 3-2. Size, Type and Use of Nails.

SIZE	LGTH (IN.)	DIAM (IN.)	REMARKS	WHERE USED
2d	1	.072	Small head	Finish work, shop work.
2d	1	.072	Large flathead	Small timber, wood shingles, lathes.
3d	1¼	.08	Small head	Finish work, shop work.
3d	1¼	.08	Large flathead	Small timber, wood shingles, lathes.
4d	1½	.098	Small head	Finish work, shop work.
4d	1½	.098	Large flathead	Small timber, lathes, shop work.
5d	1¾	.098	Small head	Finish work, shop work.
5d	1¾	.098	Large flathead	Small timber, lathes, shop work.
6d	2	.113	Small head	Finish work, casing, stops, etc., shop work.
6d	2	.113	Large flathead	Small timber, siding, sheathing, etc., shop work.
7d	2¼	.113	Small head	Casing, base, ceiling, stops, etc.
7d	2¼	.113	Large flathead	Sheathing, siding, subflooring, light framing.
8d	2½	.131	Small head	Casing, base, ceiling, wainscot, etc., shop work.
8d	2½	.131	Large flathead	Sheathing, siding, subflooring, light framing, shop work.
8d	1¼	.131	Extra-large flathead	Roll roofing, composition shingles.
9d	2¾	.131	Small head	Casing, base, ceiling, etc.
9d	2¾	.131	Large flathead	Sheathing, siding, subflooring, framing, shop work.
10d	3	.148	Small head	Casing, base, ceiling, etc., shop work.
10d	3	.148	Large flathead	Sheathing, siding, subflooring, framing, shop work.
12d	3¼	.148	Large flathead	Sheathing, subflooring, framing.
16d	3½	.162	Large flathead	Framing, bridges, etc.
20d	4	.192	Large flathead	Framing, bridges, etc.
30d	4½	.207	Large flathead	Heavy framing, bridges, etc.
40d	5	.225	Large flathead	Heavy framing, bridges, etc.
50d	5½	.244	Large flathead	Extra-heavy framing, bridges, etc.
60d	6	.262	Large flathead	Extra-heavy framing, bridges, etc.

1 This chart applies to wire nails, although it may be used to determine the length of cut nails.

57

aluminum nails are more expensive than are steel. Note that each pound of aluminum nails contains about three times as many nails as a pound of steel nails.

For fastening wood to masonry, you generally will use hardened masonry nails. If at all possible, I would recommend the use of Remington's nail driver. The nail driver fires hardened steel nails of varying lengths, with blank cartridges of one of four selections of power. It is faster, easier and neater than driving masonry nails and seems to provide greater holding power.

Nail Sizes

Nail sizes are generally designated with the term *penny*, abbreviated by using the letter d. Table 3-1 shows the approximate number of common wire nails to the pound for each generally available size. Finishing nails are lighter and will provide more, while other types will also vary a bit.

Nail costs, like everything else, have risen rapidly of late. It doesn't pay to try to economize by buying smaller nails than the job calls for. Not too many years ago, a lumber dealer would often throw in a dozen or more pounds of nails suitable to the job if you bought enough lumber or other material. As with many other things, those days are probably gone forever. If the size of the material being nailed calls for the 3½-inch length of 16 penny nails, don't expect to save money by using 3-inch long 10 penny nails, even though there are 20 more to a pound. You'll have to use more and the job probably still won't be as strong.

SCREWS

Nails provide the least holding power of all wood fasteners. When a job requires more holding power, you may have to use screws. Screws are more work to install and cost a good bit more than nails, but they provide much greater strength and, if properly driven, a neater appearance. Also, screws can be removed without damaging the work. Wood screws are commonly made of mild (unhardened) steel, aluminum, stainless steel or brass. The mild steel nails may be galvanized, chrome plated or otherwise decorated to suit the finish of the work being fastened.

Wood Screws

Common wood screws are threaded for about two-thirds of their length, starting at the tip, and have slotted heads. If you expect to drive more than a dozen screws in a day, I would recommend replacing the standard screwdriver with one of Stanley's Yankee ratcheting screwdrivers. These reversible drivers take an awful lot of the tedium out of the job, while adding only moderately to cost. They have changeable bits and can be used with just about any size of slotted screw and head design—most often Phillips or slotted. Common wood screws may come in flat head, oval head or round head styles, with either slotted heads or Phillips heads (Fig. 3-2).

Screws are best inserted by boring a pilot hole the same diameter as the screw shank in the piece of wood being fastened. Carry the pilot hole into the wood to which the first piece is being fastened. Change bits to a smaller diameter, and make sure the hole goes no further than about half the depth to be penetrated by the screw. While this sounds like a lot of work, it makes for greater ease and accuracy in placing the screws and also makes the odds on splitting the wood just about nil. If the screws are flatheads, they're meant to be countersunk. You need a countersunk bit for your drill. Oval head screws are usually countersunk, too, while round head screws are simply driven until the base of the head is firmly in touch with the work surface. The screw slots should parallel the grain of the wood (Fig. 3-3).

Lag Screws

Lag screws are also known as lag bolts and sill screws. Lag screws are much heavier than common wood screws and are often

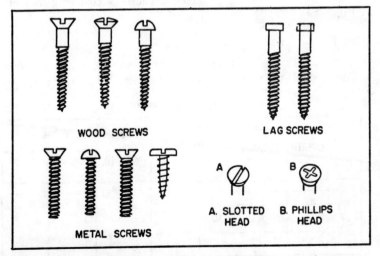

WOOD SCREWS LAG SCREWS

METAL SCREWS

A. SLOTTED HEAD B. PHILLIPS HEAD

Fig. 3-2. Types of screws.

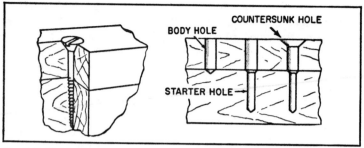

Fig. 3-3. Sinking screws correctly.

extremely helpful in attaching deck or porch ledger boards to the sill of a house. Lag screws have coarser threads than do common wood screws, and the threads seldom extend further than half the length of the screw. Heads of lag screws may be either square or hexagonal. Get the hexagonal if you intend to use a ratchet wrench and socket to drive them. Use the ratchet if at possible as it cuts way down on effort. You can buy these things in lengths up to about 16 inches. For most home work, 5 or 6 inches is about the maximum length needed, but not always. I have to pick up some 10-inch lag screws to work with a few 4 by 6s on a deck (Table 3-3).

Lag screws are also of great use in anchoring framing timbers to masonry walls. They're used with expansion anchors, which are lead inserts of a size to fit a hole drilled in the masonry and with the interior hollow slightly smaller in diameter than the lag screw being inserted. For exterior use, all screws should be galvanized, or brass, with galvanizing preferred because of cost (Fig. 3-4 and Table 3-4).

BOLTS

Bolts are useful in various types of building for two reasons. They provide the greatest strength and they also provide easy take

Table 3-3. Lag Screws.

LENGTHS (INCHES)	¼	DIAMETERS (INCHES)		
		⅜, 7/16, ½	⅝, ¾	⅞, 1
1 --------------------------------------	x	x	----------	----------
1½--------------------------------------	x	x	x	----------
2, 2½; 3, 3½; etc., 7½; 8 to 10 --	x	x	x	x
11 to 12--------------------------------	----------	x	x	x
13 to 16--------------------------------	----------	----------	x	x

60

down and reassembly. Most types of bolts, when used with wood, are used with washers to protect the wood surface, as the tightening torque can often be classified as tremendous (Fig. 3-5).

Carriage bolts are probably the most familiar form of bolt used in wood construction. The square necked carriage bolt is the most readily available. Also available are finned neck and ribbed neck carriage bolts. The reason for the neck design is simple—this is the portion of the bolt that grips the work surface as the nut is turned tight or removed. Threads generally run about three times the bolt's diameter up the shaft. Holes for carriage bolts are drilled through both work pieces. A washer should be used on the nut end, but not on the end with the bolt head. The fit should be tight so that the bolt must be driven through the hole using a hammer. You may wish to counterbore the hole enough to let the head of the bolt either sit flush on the surface of the work or under the surface. When used on wood to metal fastening, carriage bolts should always have the head end on the wood side. The square, ribbed or finned neck won't crush into the metal to allow you to turn things down tight otherwise. If you want to build bridges or other immense structures, carriage bolts are available in sizes up to 20 inches long and an inch in diameter. They also come as small as ¾-inch long and ¼-inch in diameter (Table 3-5).

Machine and *stove bolts* are also sometimes used when fastening wood. Machine bolts take the place of carriage bolts in spots where both sides of the work are easily accessible to a wrench, while flat head stove bolts can be countersunk for a slightly neater appearance (Table 3-6).

Corrugated fasteners don't have anything to do with corrugated paper boxes, though at times I wish they did. These fasteners are

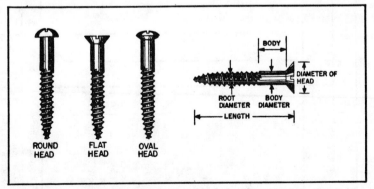

Fig. 3-4. Three types of wood screws.

Table 3-4. Screw Sizes and Dimensions.

SIZE NUMBERS

LENGTH (IN.)	0	1	2	3	4	5	6	7	8	9	10	11	12	13	14	15	16	17	18	20	22	24
1/4	x	x	x	x	x	x	x	x	x	x	x	x	x									
3/8	x	x	x	x	x	x	x	x	x	x	x	x	x		x							
1/2		x	x	x	x	x	x	x	x	x	x	x	x		x		x					
5/8		x	x	x	x	x	x	x	x	x	x	x	x		x		x		x	x		
3/4			x	x	x	x	x	x	x	x	x	x	x		x		x		x	x		
7/8			x	x	x	x	x	x	x	x	x	x	x		x		x		x	x		x
1				x	x	x	x	x	x	x	x	x	x		x		x		x	x		x
1¼					x	x	x	x	x	x	x	x	x		x		x		x	x		x
1½						x	x	x	x	x	x	x	x		x		x		x	x		x
1¾						x	x	x	x	x	x	x	x		x		x		x	x		x
2						x	x	x	x	x	x	x	x		x		x		x	x		x
2¼						x	x	x	x	x	x	x	x		x		x		x	x		x
2½							x	x	x	x	x	x	x		x		x		x	x		x
2¾							x	x	x	x	x	x	x		x		x		x	x		x
3							x	x	x	x	x	x	x		x		x		x	x		x
3½								x	x	x	x	x	x		x		x		x	x		x
4									x	x	x	x	x		x		x		x	x		x
4½										x	x	x	x		x		x		x	x		x
5												x	x		x		x		x	x		x
6													x		x		x		x	x		x
Threads per inch	32	28	26	24	22	20	18	16	15	14	13	12	11		10		9		8	8		7
Diameter of screw (in.)	.060	.073	.086	.099	.112	.125	.138	.151	.164	.177	.190	.203	.216		.242		.268		.294	.320		.372

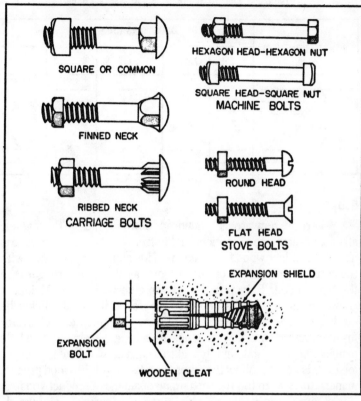

Fig. 3-5. Types of bolts.

used most often to attach mitered corners such as those in window and door molding. They are ridged pieces of metal usually 18 to 22 gauge, with one end cut off square and the other end beveled so they can be driven in. While these fasteners have greater holding power than nails and are less likely to cause splits when driven, their removal can absolutely destroy any miter joint (Fig. 3-6).

Table 3-5. Lengths and Diameters of Carriage Bolts.

LENGTHS (INCHES)	DIAMETERS (INCHES)			
	3/16, ¼, 5/16, ⅜	7/16, ½	9/16, ⅝	¾
¾	X	------	-------	-------
1	X	X	-------	-------
1¼	X	X	X	-------
1½; 2, 2½; etc., 9½; 10 to 20	X	X	X	X

Table 3-6. Lengths and Diameters of Machine Bolts.

LENGTHS (INCHES)	DIAMETERS (INCHES)				
	¼, ⅜	7/16	½, 9/16, ⅝	¾, ⅞, 1	1⅛, 1¼
¾	X	———	———	———	———
1, 1¼	X	X	X	———	———
1½; 2, 2½	X	X	X	X	———
3, 3½; 4, 4½; etc, 9½; 10 to 20.	X	X	X	X	X
21 to 25	———	———	X	X	X
26 to 39	———	———	———	X	X

GLUES

There are now probably hundreds of wood glues on the market, all making claims to exceptional holding power. It is true that glue is one of the oldest wood fastening methods and, in combination with the correct joint design, provides a joint as strong or stronger than the wood itself. Selecting the proper glue is difficult. Making a complete list for you is impossible because it would quickly be outdated by new developments. There are glues especially for furniture and cabinets. Some glues are specially formulated to hold paneling on walls indefinitely and to attach plywood subflooring to joists. It is best to follow the recommendations of the wood product manufacturer as to the type and brand of adhesive product you use. For other purposes, other gums, glues or adhesives are needed.

Animal glues are usually made from animal hides and provide a very tough wood bond, but have little in the way of water resistance. Clamping time is reasonably short at two hours for hardwood and three for softwoods.

Powdered casein is still not resistant enough to water for exterior use, but it is better than animal glues and can be used in cooler temperatures (above freezing). Clamping time is the same as for animal glues.

White liquid resin glue (polyvinyl) is still not resistant enough to moisture for weather exposure. It doesn't last as long as hide glues, but it is quick setting. This glue requiring clamping for only an hour with hardwoods and a half hour longer with softwoods.

Plastic powdered resin is highly water-resistant, though not totally waterproof. Joints must fit tightly or the joint/glue bond will be brittle. Clamping time is 16 hours for any kind of wood (oily woods such as teak cannot be glued with plastic powdered resin). A higher temperature is also needed, with a 70 degree minimum and faster setting at 90 degrees.

Aliphatic resin glues resist heat better than polyvinyl white glues, but also lack water resistance. These glues spread more easily and are less affected by some chemicals in lacquers. Clamping time is low at one hour for hardwoods and a half hour more for softwoods.

Resorcinol glues are weather-resistant. They are totally waterproof and are an ugly dark color that can stain work. The glues must be used above 70 degrees. Because they must also be mixed (3 parts of powder to 4 of a liquid catalyst) and be stain producers, resorcinol glues are usually not used unless total waterproofing is needed. Clamping time is lengthy at 16 hours.

Contact cements are used when it's time to add a plastic laminate surface such as *Formica* to your plywood table. It is brushed onto the surface and tested for dryness after 30 minutes at a 70 degree temperature. Use a piece of brown paper bag; if that doesn't

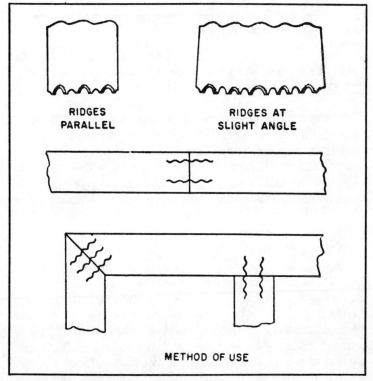

RIDGES PARALLEL

RIDGES AT SLIGHT ANGLE

METHOD OF USE

Fig. 3-6. Corrugated fasteners and their uses.

Fig. 3-7. Plain 90 degree butt joints.

stick, apply a second coat, let dry, test again and press the laminate in place. Be warned, though, that once the laminate is down it isn't going to move. The parts are then assembled. No clamping at all is needed.

Epoxy is most generally avoided when working with wood, though it will make an exceptionally strong and durable filler material and is waterproof. Some years ago I tried using epoxy as a glue for a repair project on chairs. Its almost total lack of flexibility resulted in the chairs breaking again within just a few years. Use as a filler only, or to fasten materials other than two woods.

JOINTS

While the variety of wood working joints, for carpentry, shelving or general cabinetry may seem endless, that's not quite the story. The simple butt joint is the one you'll use the most, but occasionally you'll want or need to know how to develop other types of joints (Fig. 3-7). We'll avoid covering some joints, but will still look at most. Sooner or later you'll probably need to join a couple of boards, cross one shelf over another, brace under a shelf or apply molding to a door or window. You may even wish to change a design here and there and use a mortise and tenon joint to brace a picnic table's legs or miter a corner joint on a planter.

Generally, joints are strengthened using other fasteners—whether glue, nails, screws or dowels and glue. For fancier joints you'll wish to add a good set of chisels to your backsaw, and you may even want, a router to make some of the cuts.

Half-Lap Joints

Half-lap joints are used to join pieces of the same thickness. A simple lap joint has two boards lapped and fastened, while a half-lap joint has half of each board cut through on the backsaw and then fastened. For corner half-laps, you measure off the needed amount of lap from the end of each piece and square a line all the way around. Pick the best or strongest looking surface of each piece and place that face up. Then score a line at the halfway mark for the thickness of each piece. If you use a marking gauge, score both pieces from their faces. If you miss a bit and the distance is not exactly half on one, the material remaining on the other will make up for it. Make the cuts on your original line, down to the gauged lines (sawing from

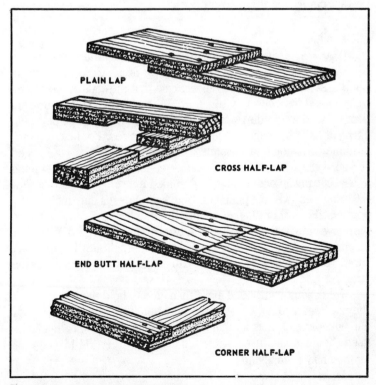

PLAIN LAP

CROSS HALF-LAP

END BUTT HALF-LAP

CORNER HALF-LAP

Fig. 3-8. Some types of lap joints.

the back of the first piece and the face of the other). Since end laps can be cut with a saw, use a fine toothed ripsaw. Or carefully use a crosscut saw to cut along the gauge lines until they merge with the cuts already made (Fig. 3-18).

For cross half-laps, as shown in Fig. 3-8, you figure the lapping point and lay the boards across each other and mark the intersections. You'll have to turn them over to mark both. Square the lines and then figure your depth, again depending on the half-thickness but marking from the face side of each piece in order to prevent problems. Cut along the square lines down to the gauged depth. Next, you'll need a good, sharp chisel. The chiseling will be easier if you use the backsaw to cut a series of saw kerfs ¼ inch or so apart, down to the marked depth, across each section to be removed.

Start chiseling slightly above the marked line for your rough cut, holding the chisel with its beveled edge up. Then use a paring chisel to finish the joint bottoms. This can be done more quickly and, with experience and the right bits, more accurately with a power router. Fit the pieces together. Use the fasteners of your choice.

Miter Joints

You'll need *miter joints* when building outdoor furniture, decks or anything that may need decorative or finish trim. Mitering is nothing more than cutting at an angle so that the ends of the pieces to be joined fit accurately together. Most often you'll be mitering door or window moldings. Since the final angle must be 90 degrees, each cut must be made at half that or 45 degrees. For fancier stuff, assuming you don't have Stanley's new mid-range box, you'll need to figure just what the angles need to be for each shape. Assuming you're cutting an octagon (an eight-sided figure), you make cuts at 67½ degrees, while a pentagon (five sides) needs cuts made at 54 degrees. For other shapes, divide the number of sides the figure will have into 180 degrees and subtract the answer from 90. For example, a 10-sided figure will work out in the following manner. First, 180 divided by 10 equals 18; 90 minus 18 is 72 degrees for the cuts.

Miter joints will need more than glue to hold them together if they're freestanding. Most of the time you'll be working with door and window moldings so that a few brads or finishing nails can be used. You can also use corrugated fasteners (Fig. 3-9).

Mortise and Tenon Joints

Mortise and *tenon joints* are great for making strong outdoor furniture, but require a fair amount of work. Most people don't

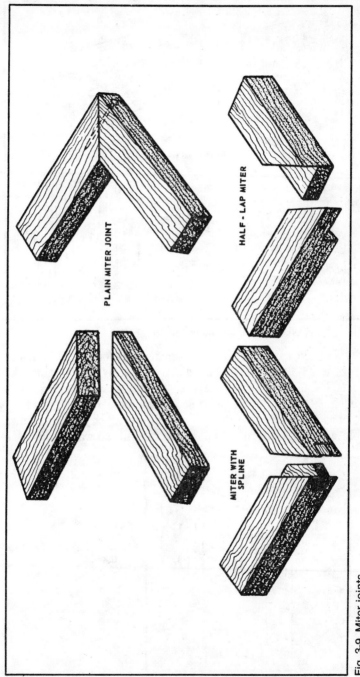

PLAIN MITER JOINT

HALF - LAP MITER

MITER WITH SPLINE

Fig. 3-9. Miter joints.

69

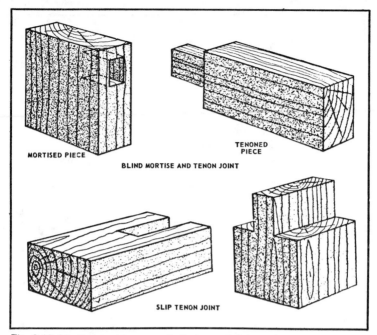

Fig. 3-10. Mortise and tenon and slip-tenon joints.

bother, instead preferring to use simpler joint forms requiring less work which have to be fastened more securely and don't look as nice. The tenon is usually fairly easy to cut, while the mortise will require routing or chiseling and a fair amount of care to get a tight joint and good strength. A blind mortise and tenon joint is simply a joint where the tenon does not penetrate all the way through the

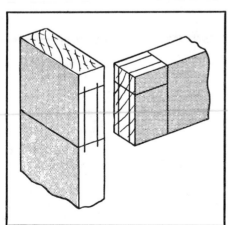

Fig. 3-11. Layout of a stub mortise and tenon joint.

Fig. 3-12. Dado and gain joints.

mortised piece. A through mortise and tenon allows the tenon to come to the opposite side of the mortised piece. The through mortise and tenon is easier to make.

Usually the tenon will be one-third to one-half the thickness of the piece to be mortised (the thicker that piece, the closer to one-half you should come). A mortising gauge can be used to mark the tenon but is not essential. Determine the size, both length and width, of the tenon and mark the work piece carefully. Use a backsaw to cut the tenon, preferably in a miter box for accuracy. Mark the mortise in the same manner, and use a drill to remove as much of the material inside the marked lines as you can. Finish up

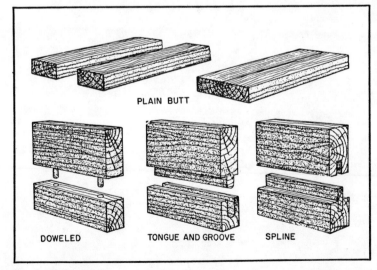

Fig. 3-13. Edge joints.

with a mortising chisel. Blind mortise and tenon joints require the use of a depth gauge on the drill. You can run the drill bit through a wooden block and set it at the correct depth if you don't have a depth gauge. Or, you can take several strips of a bright colored tape and mark the bit at the correct spot. Just be very careful not to penetrate beyond the marks.

Mortise and tenon joints are usually glued. For outdoor use, a resorcinol glue should be used. The tighter the fit with this glue, the better, so a lot of care is needed when making the joint. Once glued, the mortise and tenon joint should last as well as just about any possible method of fastening, and it will look very neat and professional. (Figs. 3-10 and 3-11).

There are quite a few other joints, but I've covered only the ones most useful in outdoor construction. For details on others, almost any book on cabinetmaking is a good place to look (Figs. 3-12 and 3-13).

Chapter 4
Foundations

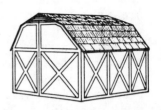

Foundation means base. Without a good base, no building, deck or other piece of work will last very long. In some cases, outdoor buildings are so small that they can shift with frost heave with little or no harm. When size goes much beyond 8 feet by 10 feet, some sort of solid base, reaching below frost line, is essential.

Material, assuming it is impervious to rot, is of less importance than strength and correct depth of setting. Unfortunately, I can't begin to list the depths needed here for the area in which you may live, as these vary widely depending on local conditions. My present locale has a frost depth of only about 1 foot, while my former home had a frost depth of 42 inches. In certain areas of the country, it is virtually impossible to build on anything other than piers because of high water levels. The variety of soils, each with a different bearing capacity, will affect depth and often footing width.

Wall foundations are solid, running under the entire length and across the entire width of the building. They are best for support when you expect the loading to be heavy. Materials may vary from poured concrete to concrete block and stone. Masonry wall footings follow a rule of thumb requiring them to be twice the width of the foundation wall in cross section and as deep as the foundation wall is wide.

Pier foundations can save time and labor and may be of masonry or pressure-treated wood. Spacing varies according to weight loads. In most instances, on center distances for piers will be no less than 6 feet and no more than 10 feet. The use of piers as foundations for a building will require heavier sill boards than will a solid wall. Any extra cost here is easily made up by the savings in excavation and materials for the piers.

SELECTING A SITE

Before starting a foundation for even the smallest tool shed, you need to select an appropriate site and then make the excavations. When choosing a site, avoid rock ledge formations whenever possible. Place the building as you wish, of course, but consider a few things such as sun angle, prevailing winds and any need for light. If the building is to be heated, placing it so that any windows face south or southeast is a good idea. If you're wiring the building, whether or not the windows provide light for a good portion of the day is of less importance. With today's outrageous electrical costs, it should still be a consideration.

Workshops and tool sheds placed too close to other buildings can make the handling of large tools and materials more difficult than is necessary. Any structure placed too close to a property line may violate local codes. Check your local codes carefully, for if you don't have the proper setback you can be compelled to move or raze any building. Walls, decks and other structures may also be subject to code requirements regarding structural rigidity, overall strength and placement. Sometimes even the appearance of a wall or out-building will affect the chances you have of getting a building permit. Sometimes you will have to show plans; others times a rough drawing or a verbal description will suffice.

LAYOUT

Once the codes are understood and met, you've reached the point where laying out the foundation is necessary. The job must be

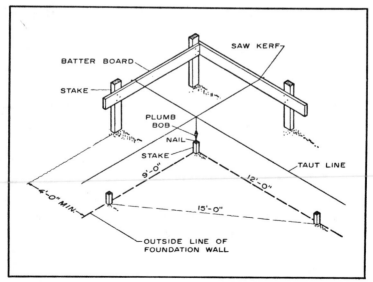

Fig. 4-1. Corner setup for house layout.

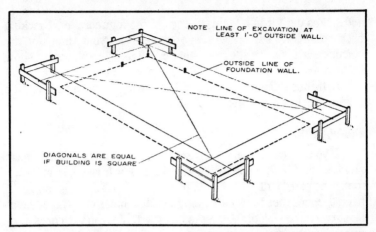

NOTE LINE OF EXCAVATION AT LEAST 1'-0" OUTSIDE WALL.

OUTSIDE LINE OF FOUNDATION WALL.

DIAGONALS ARE EQUAL IF BUILDING IS SQUARE

Fig. 4-2. Overall layout of house or building set and ready to go. Make certain the diagonals are exactly equal.

done accurately, but is not as hard as it first appears. You'll be working from some sort of a plan no matter what type of structure is going up, so you'll know what the exterior dimensions will be. We'll work here on the basis of a simple rectangle. If your building is to be L-shaped, you need only work as if you have two simple rectangles joined along one section.

Determine where you wish to place the first corner of the outbuilding. Drive a small nail into the top of the stake. Approximately 4 feet outside of this stake, make a setup, as shown in Fig. 4-1 of stakes and batter boards. Saw kerfs are cut in the batter boards at the point where lines drawn from the stake will intercept them, if you have a 90 degree angle from the stake. Measure the distance needed, from the first stake (with the nail in it), for the exterior wall. Drive another small stake and pound a nail into its top. Again, place stakes and batter boards. From this point, measure the needed distances to the other corners, drive the stakes and place batter boards.

It's at this point that you must begin to worry about corner angle accuracy. Run mason's line from each saw kerf on each batter board to the next in line. Place a light plumb bob at the point where the lines intersect as they leave the batter boards. The plumb bob should drop directly to the nail head in the center of your small stake, as you see in Fig. 4-1. Now measure up your line on one side 9 feet, and 12 feet up the other. From these two points, a measurement across will be 15 feet if the corner is exactly 90 degrees. If the measurement isn't correct, move the batter boards and stakes until it is. Do this at each of the four corners, checking back on already set corners as you go. Take your time, for the correct angle is critical to a straight building. Once all angles show 90 degrees, measure the diagonals, from one corner stake to another. If these diagonals are equal, your layout is square and it's time to begin excavation (Fig. 4-2).

EXCAVATION

If you are using pressure-treated 4 by 6s as piers, excavation is simple. Just use a posthole digger, going below frost depth at least 1 foot. The piers will probably be placed on 6-foot centers, so you'll need to mark these off before digging. The holes dug should be at least twice the size of the post to be inserted. Even with pressure-treated lumber used for piers, it is a good idea to apply a preservative to the cut ends if you make cuts. If possible, avoid placing the cut ends below ground.

For masonry piers, the size of the hole dug will depend on the materials used. If a poured footing is to be used under a masonry pier, simply dig the hole the size of the footing, and pour the footing. Greatest strength for poured footings joined to poured piers is achieved if you insert at least one steel reinforcing rod in each

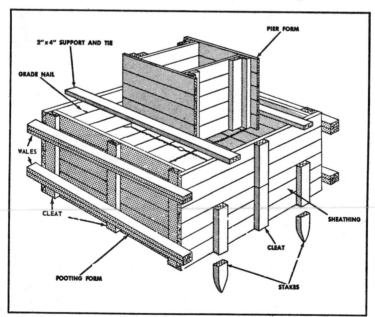

Fig. 4-3. Footing and pier form combined.

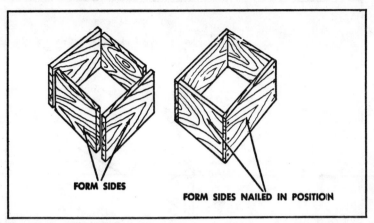

FORM SIDES

FORM SIDES NAILED IN POSITION

Fig. 4-4. Simple footing form.

footing so that the poured pier will be well joined to the footing. If the footing is to be laid up, either concrete block or stone, such a reinforcing rod will also add strength there. For laid up masonry footings, the holes will need to be large enough to allow you to work in them to lay the block or stone. You can also buy precast concrete piers and footings which can just be set into the proper sized hole. Pier footings should be at least 2 feet square and no less than 8 inches deep, with a foot preferable if the building is of any size at all.

Poured concrete footings and piers can be done at the same time if you make forms for the purpose. Simply provide a form with the correct interior dimensions for the footing. Add one with the proper dimensions for the pier to that, using a couple of cross boards to support the pier form (Fig. 4-3). Because masonry footings are usually larger than wood ones, you will probably use fewer of them by placing them on 8 or 10-foot centers (Fig. 4-4).

All footings, whether for piers or foundation walls, must be placed on undisturbed ground. In other words, don't dig past the depth needed for the footing and then try to refill the hole with dirt or gravel. If you go too far down, remove the loose soil and make the footing deeper at that particular point.

Making Continuous Footings

Continuous footings are needed for foundation walls. We've already covered footing dimensions, so the method of making them is all we need now. In most cases, whether you're laying a slab floor for a garage, placing a full basement, or getting ready to use a crawl space, the footing will be of poured concrete. This is strongest and

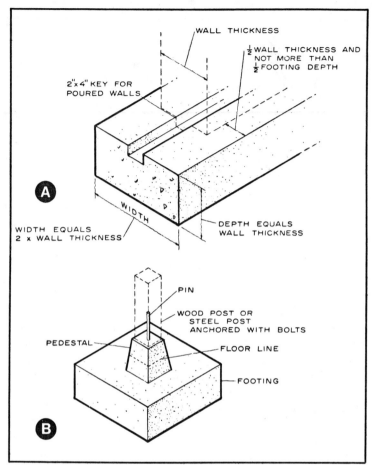

Fig. 4-5. Concrete footing. (A). For a poured wall foundation, the keyway is made by laying a 2 by 4 in the wall after the cement is poured. The keyway is not needed if concrete block walls are used for the foundation. (B). A post footing can often be bought precast.

can be aided by laying an oiled 2 by 4 down the middle of the footing (if the foundation wall is to be poured) to form a key to help prevent shifting and water entry (Figs. 4-5A and 4-5B). Footings can be poured in a sharply cut trench or into forms. A bit of care in digging the trench can save quite a few bucks as rough lumber for forms costs quite a bit of money these days. Make the general excavation for a full or partial basement as rough as needed, making sure there is enough room to the outside of the wall placement to allow you to work. Use care when digging the footing trench.

Basement

You may find it desirable to provide a full basement for an outbuilding. If that's the case, make sure the resulting wall height will allow an interior ceiling height of at least 7 feet 4 inches. Give some thought to running at least one extra course of block to provide an 8-foot high ceiling. If, as often happens in residences, heating runs, plumbing runs and wiring will pass through, give consideration to another course as well so that there will be plenty of height downstairs.

Footings and foundations for exterior walls of brick follow the same rules as for residences. Some forms of loose stone walls are simply laid on top of the ground. These require no footings since frost heave is usually absorbed with little effect, but any wall erected with mortar must have a footing to prevent joint cracking (Figs. 4-6 and 4-7).

CONCRETE

Concrete is frequently used in outdoor building projects, and this is as good a time as any to cover some of its properties. Portland cement is the type normally used, and various mix proportions provide differing strength characteristics. Sand and any gravel

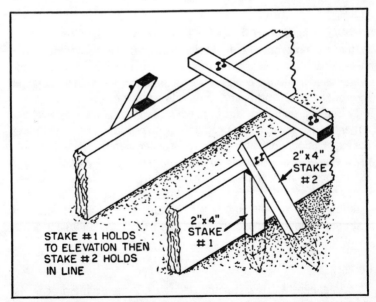

Fig. 4-6. If the footing trench is not sharply cut, wall footing forms such as these must be used.

aggregate should be clean. In most cases, aggregate should be available in sizes ranging from ¼ inch to no more than 1½ inches. However, in a few areas of the country you won't be able to locate anything that large. Then it is a good idea to increase the amount of cement used to make the mix by about 5 percent if the largest stones in the gravel are 1 inch in diameter. Use 10 percent more mix if the largest stones are only ¾ inch in size. For stone where the maximum available size is no more than ⅜ inch, a full 20 percent increase is needed.

Most of the concrete you use will probably be mixed at home. Outbuildings are seldom of a size to make it sensible to have a ready-mix truck deliver four or five cubic yards of concrete.

To determine the amount of concrete needed, simply take the dimensions of the project—length, width and height—and figure the area and then the volume of the project in square feet. Convert to square yards. A wall of concrete 9 feet long and 3 feet high by 1 foot thick would give you 27 cubic feet (9 times 3 times 1). Divide that by 27 (the number of cubic feet in a cubic yard) and you get the volume of concrete needed for that particular job. I sneakily selected a simple example, but follow the formula and you'll easily be able to estimate you needs.

Water

Water is of great importance when mixing concrete. Without it, there is no process of hydration (the chemical hardening of the mix). With too much or the wrong kind, the mix will be weakened. Water used in mixing concrete should be clean, fit for drinking, and free of sediment and excess minerals whenever possible.

Use just enough water to make tooling the concrete easy, but not enough to make it sloppy. There is no possible way to estimate the exact amount of water going into a concrete mix, since there are great variables of wetness in the various aggregates you'll be using. If the coarse aggregate has been thoroughly wetted down, less water will be needed to get a correct mix. Even less will be needed if the fine aggregate (sand) is very damp or wet. Cure strength of concrete is most often expressed in seven day rates, though in some cases a 28 day rate may be used.

Curing

As concrete dries, it gains strength—if the drying process is controlled. Too rapid drying doesn't allow the hydration process a chance to get near completion, so that overall strength is greatly

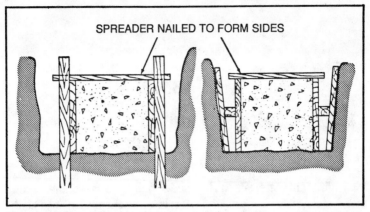

SPREADER NAILED TO FORM SIDES

Fig. 4-7. Footing forms must be braced.

reduced. At a minimum, concrete should be kept damp for at least four days, with seven preferable. In most small construction, a 28 day curing time is simply not practical since total construction time may be well under two weeks. To properly cure flat work such as a slab or floor, you can cover it with burlap, straw or other material. These days polyethylene sheeting is considered the best. The sheeting is impervious to water, so that further wetting of the slab is seldom required. Using other materials, a daily sprinkling in dry weather will probably be essential. For a good cure, wet the surface of the slab well and immediately cover. If the material used for a cover is not waterproof, prevent evaportaion by checking daily for dampness and wet down if near dry.

Walls poured with forms are easily kept damp by a daily wetting of the wood in the forms. Soak the wood thoroughly. Cover the tops of the walls with any of the materials you might use to help cure a slab.

The thinner the concrete in use, the more important the job of curing is to final strength. Since poured concrete may prove useful for many outdoor projects, the thickness will surely vary. It is even possible to cast concrete bridges, though not many people wish to go to the expense, work and bother. A wooden bridge or even a precast concrete culvert pipe will usually do the same job without requiring (assuming a 6 ton load and a 6-foot span) a 6½-inch thick slab, which would need ⅝-inch reinforcing rods spaced every 8 inches.

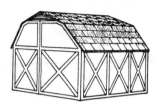

Chapter 5
Fences

Many different materials can be used to build fences—stone, wood, metal and brick. The major factor in any fencing job is usually durability, whether the fence is designed to hold livestock, keep out intruders, or simply to enhance the appearance of a home or farm. Of course, sometimes fence longevity is not essential as in stock fences around areas being grazed for only a short period of time. Farmers will usually erect a single or double strand electric fence to contain the livestock, using fiber glass or metal posts to hold the electrified wires at the appropriate height. Such fences go up and come down easily and quickly.

In most cases, though, fences are intended as permanent or semi-permanent structures and need to be designed to withstand wind loads, weathering, post rot and animals. Materials selection can be critical as a poorly chosen wood will rot, warp, or be easily knocked from its moorings. Even decorative fence design can become critical, as these fences must withstand strong winds. Good fence post setting precedures will win much of the durability battle, but first you need to select materials with reasonable care.

WOOD FENCING MATERIALS

Wood fencing materials include cedars, various grades of redwood, pressure-treated pine and durable hardwoods. Cypress would be another possibility, but it has limited availability in most areas and tends to be very expensive. Of the hardwoods, white oak is the most durable of the oaks, followed by live oak. Post locust will about double the life expectancy, in contact with the ground, of white oak. Such woods are difficult to work with and are costly to replace. In areas of great dampness with a lot of insect activity, no hardwood lasts well in ground contact. White and live oak would remain good woods for plank fencing. Drill before nailing to prevent splintering, especially when working close to the ends of boards.

Redwood and much cedar is cut on the West Coast, with cedar also cut extensively in the upper Northeast. Transportation costs can add heavily to expenses when using large amounts of lumber. This added cost is sure to rise as fuel costs increase. Still, both redwood and cedar are extremely attractive woods and easily worked. I feel that redwood has a slight advantage here as it isn't quite as brittle and likely to split as the cedars. Neither wood is as strong as pine and fir, though, so the greatest use would be for decorative fencing. Cost could easily be the deciding factor. Pressure-treated pine goes for about half the price of similar quality redwood in the East. My brother informed me that redwood in California sold for little more than treated pines, so the transport really does make a difference (at present rates, possibly as much as $900 per thousand board feet).

Pines and firs are not durable in contact with the ground. As far as I can find out, no surface method of treating these woods adds significantly to in-ground life. But pressure treating with chemicals such as those provided by Koppers Company and its Wolmanized and Outdoor lumber products does. After three and a half decades, many examples show no signs of rot. Some early wood treatments used around the London docks showed no signs of rot after more than 100 years.

One scale shows all cedars, cypress and redwood with a grading of one (the best) in decay resistance, with Southern yellow pine rated at two. Only pitch pine among pines and firs manages to finish as high as three (on a scale of one to five). Chestnut hits one among the hardwoods; locust is rated at 2.

Constant exposure to dampness is a strong factor in wood decay and rot. In most instances where fences are to be of any length, I would recommend pressure-treated pine for economy, ease of workability, strength and durability. The economy is aided by the fact that locally cut species can be used, so only the chemicals need to be shipped.

METAL FENCES

Metal fences may come in more types and styles than do wood fences. Metal fences are less likely to be used for decorative purposes in most areas as the material is generally less attractive than wood and often more expensive. In some cases, metal fencing requires more skill and more tools to erect. Barbed wire and diamond mesh wire needs special stretchers, as do most of the meshes suitable for equine use.

Chain Link Fencing

Metal fencing should be either galvanized or aluminum for greatest durability. Some companies, including Sears, now offer galvanized *chain link fencing* that also has a polymer (plastic) coating. Chain link fencing isn't cheap, but it is excellent material. Chain link in lower—3 to 6 foot—heights is excellent for protecting yards from trespassers and helps keep tots out of traffic. For smaller dogs, the 4 to 6-foot heights are excellent for keeping them on your property, or for making runs for kennels. For larger dogs, you will probably have to add a mesh roofing of some kind. I've seen medium sized large dogs go over a 6-foot chain link fence almost as if it wasn't there. The open links on industrial height fences provide enough paw or claw hold for large dogs to scramble over even 10 or 12-foot fences.

For good service, chain link mesh should be no less than 12 gauge wire, with 9 gauge used for extra strength. The lower the gauge number, the thicker the wire. Posts will have an outside diameter of from 1⅜ inches to as much as 2½ inches. For residential use, companies such as Sears and Montgomery Ward offer outfits that include the fabric, line posts which should be spaced no more than 10 feet apart, loop caps to hold the top rails, top rails and fittings for end posts, and mesh holders and tie wires to hold the fabric to the top rails.

End posts are not included because the shape of each fencing job will change the number, so these are ordered separately. Usually the end posts will include tension straps, and caps.

You can also buy already assembled chain link panels as large as 10 by 6 feet, though the cost is fairly high. These panels offer the advantage of exceptionally easy erection and ease of mobility, if one wishes to move the fenced area from year to year or month to month.

Stock and Poultry Fencing

Stock and *poultry fencing* comes in so many styles and strengths that only a general coverage is really possible. Steel netting, or chicken wire, with a 2-inch mesh is almost ideal for enclosing garden spaces to keep out rabbits and other small animals. For garden use, I would recommend a mesh no less than 3 feet high, with some preference for 4 feet. Two-inch mesh is sufficient most of the time, as is 20 gauge weight (more or less standard).

Stock fence for cattle, hogs and horses is a different matter. Heavier wire is used; Sears offers theirs in ¼-inch diameter. Fences for hogs must extend at least a small distance below ground to be really effective. Cattle are less prone to try and get under a fence, but they may lean against it to scratch; thus, posts must be sturdy and fabric should be strong. Generally, a 6 by 8-inch mesh is sufficient to hold cattle. For hogs you will need a tighter lower mesh, possibly as tight as 2 by 8 inches. Many farmers prefer to get the cheaper open mesh, bury a foot or so of it and use a bottom board along the base of the fence to keep the hogs from separating the mesh. Hogs are among the hardest animals to keep penned, so a sturdy, tight fence is essential.

Equine fencing is a touchy subject. In many areas of the country, barbed wire is still used. But its use is declining, as it should. Horses are physiologically and psychologically equipped to flee when spotting anything they perceive as dangerous. I've seen horses bolt at a falling leaf. Electric fencing can be good for unspirited horses, but it shouldn't be used for stallions. Stallions are often edgier than mares and geldings and will sometimes bolt through an electric fence before they really know they're being shocked.

Mesh fencing presents another problem. The mesh must be tight enough so that the horse cannot catch a hoof in it. For most adult horses, a mesh no larger than about 4 by 5 inches is needed, while foals will require smaller meshes. Horses panic easily when a hoof is confined and can severely injure themselves if hung up in a fence. This is why barbed wire is not a good holding device for horses. Range horses are less likely to get hung up on barbed wire, probably because they are more used to it.

Horse eyesight can be a bit strange at times, and wire fencing should always be supplemented with some visual clues, especially if electric fence is used. That single thin strand of wire can remain unseen by the horse, which may simply, walk or run through the barrier. Basically, all equine fencing of metal wire, or mesh, needs some visual support. The thinner the wire or mesh, the greater the need, but even fairly heavy gauge mesh can drop out of equine eyesight against a rising or setting sun. Most mesh fencing will do fine with a 1 by 6 board along the top of the fence line, while electric fence can use strips of rag or old commercial pie plates.

ELECTRIC FENCES

One of the cheapest fences for stock holding is the *electric fence*. Battery-powered fence chargers can be bought for as little as $20 or so, and will charge up to 10 miles of single wire fence. Good, transistorized plug-in chargers cost under $50. These will handle up to 20 miles of fence. Sears is now selling a solar-powered fence charger for less than $150. The unit uses sun power during the day and a 6-volt battery at night. The battery automatically recharges during the day and has, according to Sears, a three week reserve of power should the sun fail to shine for that long.

Battery-powered fence chargers are probably best left for remote areas where other power is unavailable. The expense and time spent changing batteries can be greater than the cost of electricity for a plug-in style. As for the solar-powered version, I simply don't know if it is worth three times the cost of a plug-in or battery style. Sears doesn't say whether or not the rechargeable battery is replaceable. It has been my experience with rechargeable nickel cadmium batteries that maximum life is about five years. I can't be sure, though, of the life of this battery as Sears calls it a gel type.

Posts for electric fence are much lighter than for other types of fencing. I used 4-foot fiber glass posts and a few scrub trees with insulators nailed to lath, which was then nailed to the tree to keep the tree from growing out over the insulator. These posts cost a bit under a dollar apiece and can be placed 12 to 15 feet apart. Insulators are relatively inexpensive if steel or wood posts are used, and "gate" fixtures are under $1.50.

Wire used for electric fence will be lighter than the usual wire for fencing. Usually, 17 gauge wire will do. Aluminum wire can also be used and will do a better job of conducting electricity than steel, though the wire costs a dollar or two more for a roll of 1,320 feet for about the same gauge (15 gauge aluminum, 14 gauge steel).

BARBED WIRE

Barbed wire comes in many configurations, although most of what I see today is the two point steel kind, sometimes with copper plating on the steel points. Lightweight barbed wire (18 gauge) is used in conjunction with electric fence chargers, while middleweight (14-16 gauge) and heavyweight (12 gauge) is used alone. Almost all barbed wire is used for cattle or sheep fencing. Cattle are less susceptible to panic than horses, have somewhat tougher hides and are characteristically less active. Sheep probably shouldn't be contained with barbed wire, for they are even more susceptible to panic than horses.

FENCE ERECTION PRINCIPLES

Just following the principles of good carpentry can provide an excellent and durable fence. There are a few special rules to consider, but the basics are the same as for almost any kind of construction. If the underpinnings aren't sturdy, the fence will be flimsy, no matter how heavily constructed above ground. The start of the job depends on the location. If you live on a square or rectangular city or suburban lot, you may need do nothing more than locate the corner posts and measure for your postholes. From there, it is a simple matter to get the proper depth postholes and go on. If you live in a rural area, you may need to lay out a fence line and then clear the line of brush, trees, stumps, rocks and old fence.

Layout

Layout starts with two people. First, locate the proposed ends of one section of fence and set a stake at those points. One person stands at one end, while the second carries a pole down the line (Fig. 5-1). The first person checks the lineup of the three stakes, and another is driven at the point checked. Either the end post or the check post (or stake) can be moved if needed to get correct alignment. Continue this process, working at about 100-foot intervals or less, until the one side is laid out. Then move on to other sides.

If you have to go over a hill or rise where the end stakes cannot be seen, set two stakes at the top of the rise so that both can be seen from each end stake. These stakes are then aligned with the end

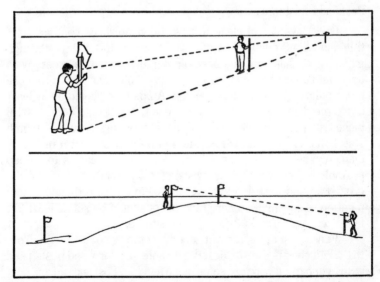

Fig. 5-1. Laying out a fence line. The top illustration shows the process being done on level or nearly level ground. The bottom drawing shows the way to go on hilly terrain.

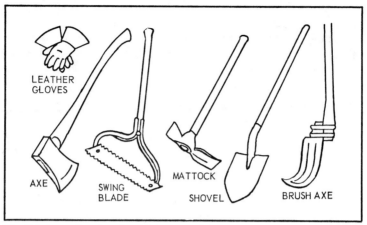

Fig. 5-2. Some hand tools which are useful in clearing brush from a fence line.

stakes. This process may take some time, as you'll often find it necessary to move the intermediate stakes several times to get good alignment.

Once the fence lines are laid out, you may need to start clearing for the fence. Trees are best laid down with a chain saw, while the tools in Fig. 5-2 are good for clearing smaller obstacles.

Posts and Postholes

Once layout and clearing have been finished, you can start thinking about posthole depth. I realize, unfortunately through experience, that the hand posthole digger just isn't a great joy to use. But undersized and too shallow postholes contribute to more fence failures than any other factor. A posthole should be at least twice the diameter of the post being set, and should be at least 6 inches deeper than the post will actually be set. A rule of thumb states that anchor and gate posts will be no less than 5 inches in diameter, set no less than 3½ feet deep. That means 4-foot deep postholes. You have my best wishes for dry, sandy ground, for the ground around here is often wet and heavy with red clay, not to mention many small stones. Line posts should be set at least 2½ feet deep, requiring a 3-foot deep posthole.

Fence post diameter will depend on the use of the fence. Note that 2½ inches is the smallest reasonable size for wood posts used with light barbed wire and some mesh fencing. Post and rail and post and board fence will require heavier posts along the fence line, usually 4 to 5 inches. Anchor posts need to be at least twice the

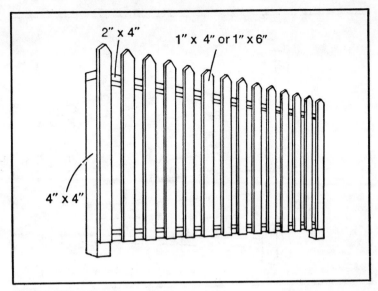

Fig. 5-3. Picket fence.

diameter of line posts. Anchor posts include corner posts, end posts, gate posts and any braced line posts. Fence height is again a function of the fence use. To hold a stallion, a minimum of 5½ feet is recommended, though some people settle for 5 feet, while a bull will be easily held with a 4-foot high fence.

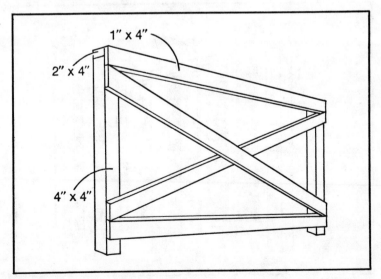

Fig. 5-4. Post and rail fence.

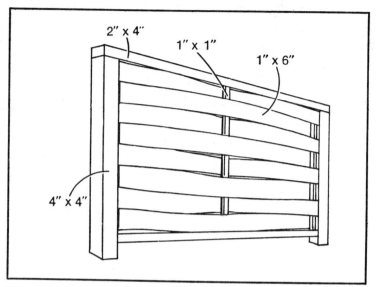

Fig. 5-5. Basketweave fence.

Decorative fencing can be about any height you care to make, though screen fencing designed to provide privacy or hide unattractive areas is seldom less than 6 feet tall. Many decorative border fences are no more than 2½ feet high (Figs. 5-3 through 5-14).

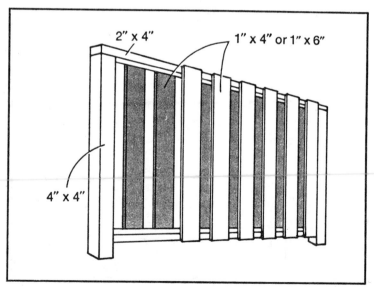

Fig. 5-6. Board and board fence.

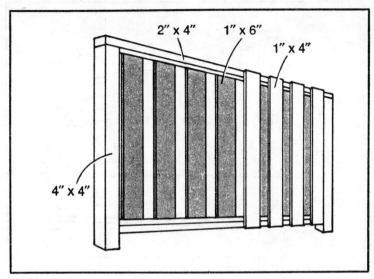

Fig. 5-7. A fence with boards of alternate widths.

Garden fencing should be at least 3 feet high and seldom needs to be more than 4 feet high. If you have problems with deer getting into your garden, building a fence high enough to keep them out could kill all the savings expected from a decade of growing your own vegetables.

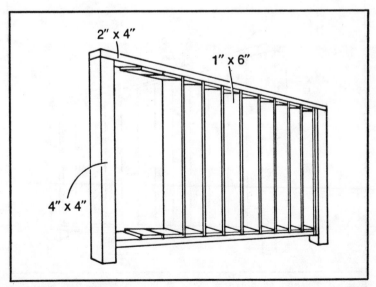

Fig. 5-8. Louver fence.

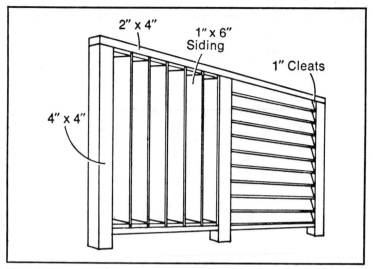

Fig. 5-9. Fence with alternate louvers.

Spot posthole positions using small stakes and start digging (Fig. 5-15). As mentioned earlier, the postholes are dug twice the diameter and 6 inches deeper than the post will actually be buried. Fill the 6-inch space with gravel or small stone to aid drainage. You can skip the extra 6-inch depth if you live in a very dry area or in an

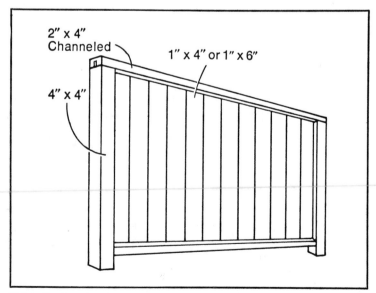

Fig. 5-10. Channel panel fence.

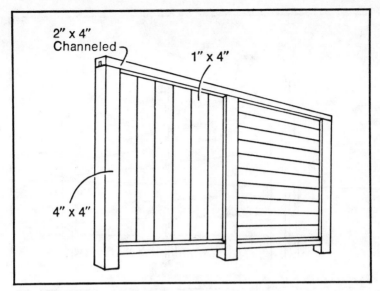

Fig. 5-11. A fence with vertical and horizontal panels.

area with well-drained soil. There are several methods of making sure the posts won't shift after they are in the ground. Use of concrete for a foot or so up the post. If concrete is used to secure the posts, lay a flat stone over the gravel at the base to keep the concrete from interfering with drainage. Otherwise, it is possible to

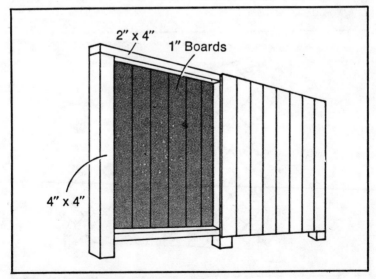

Fig. 5-12. A fence with alternate panels.

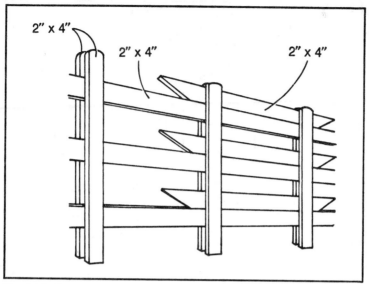

Fig. 5-13. Shaped ends fence.

just tamp down the dirt. You can also use a wood section forming a "T" at the base of the post. For this, you must open up the posthole and then nail a pressure-treated board three times the width of the post across it. Use galvanized or aluminum nails at least three times the thickness of the board being nailed. Fill the hole and proceed (Fig. 5-16).

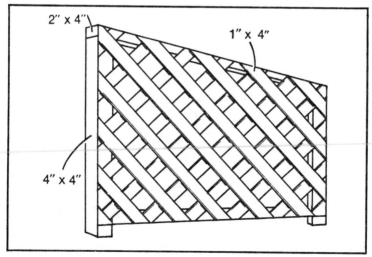

Fig. 5-14. Trellis fence.

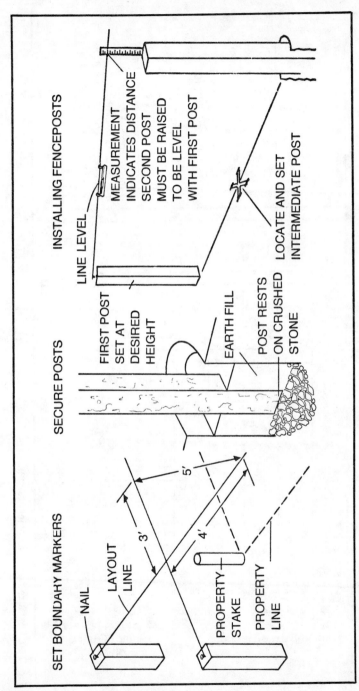

Fig. 5-15. Proper methods for spotting, securing and installing fence posts.

95

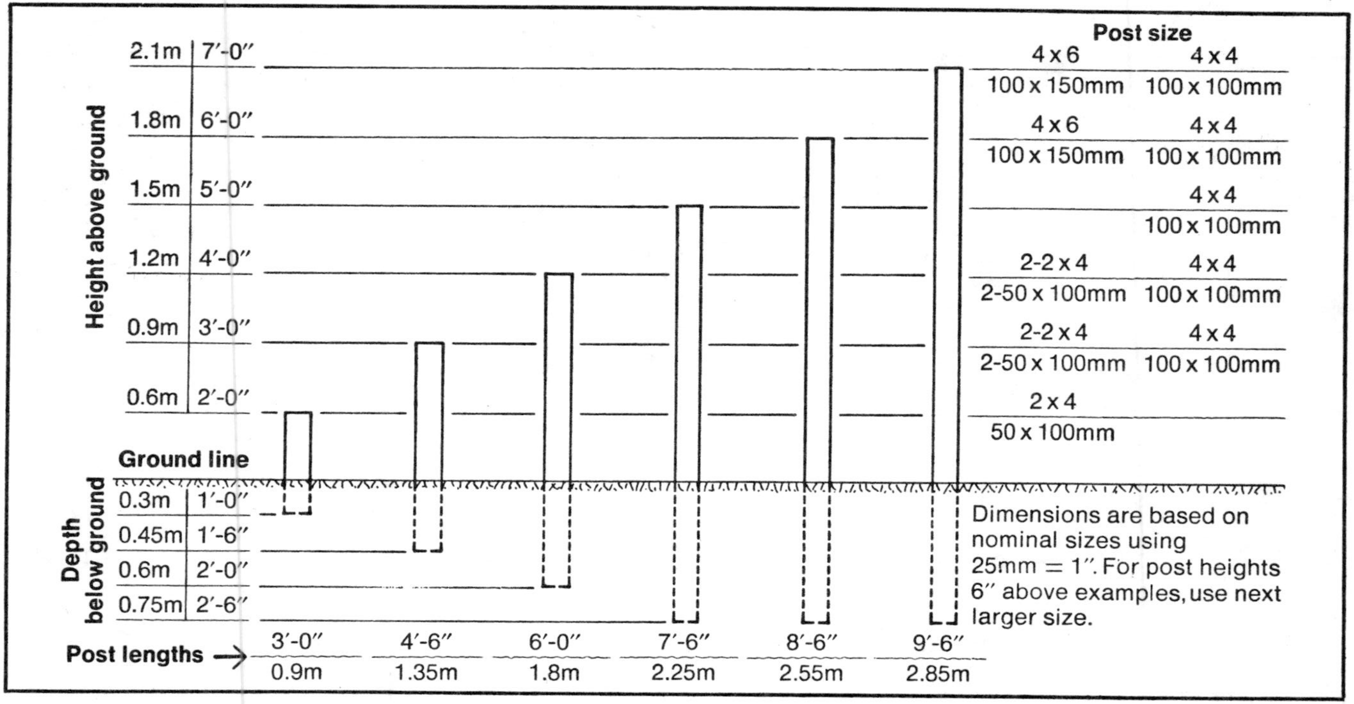

Fig. 5-16. Post size and embedment depths.

Erecting Board Fences

If you are building a *board fence*, erecting all the posts at one time will speed the job. For post and rail fencing, where the rails fit into holes bored in the posts, you can't put up more than a single fully set post at a time. Set the first post securely and place the second. Insert the rails into the bored holes in the well-set post. Then tilt the second post back slightly to receive the other ends of the rails. This is much more easily done with two people, as the top rail often has tendency to slip out as you get ready to move the post into its upright position. Set the second post, move on to the third and so on around your perimeter. Each post should be plumbed on two sides, as well as is possible. The accuracy of the plumbing job is affected by the type of post used. Smooth posts can be dead accurate, but rustic types will probably be only close.

Board fences can be built in many designs. In Fig. 5-17 you see two popular designs. The standard four board fence is usually made up of 1-inch pine boards on 4 by 4 posts. The boards should be pine because oak, even though stronger, has a tendency to break in spear-like shapes when struck hard.

For feedlot use, or in any area where livestock might be closely confined in space, the fence will need to be heavier. Instead of 1 by 6 boards, 2 by 8s need to be used. For very heavy use, 6 by 6 posts on 6-foot centers are a good idea. Use annular or screw shank nails at least three times the board's width, in nail length, for good holding power; use no less than three nails to each end of the board being secured. Boards twice the length of a single panel are always preferred as they add stiffness to the overall fence, especially if you work so as to end no more than two boards on any one post. Boards on fences used to hold livestock are fastened to the inside, or livestock side, of the posts. Decorative board fences may have the

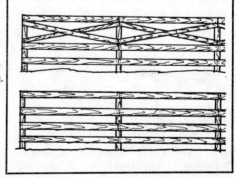

Fig. 5-17. Two common designs for board fencing.

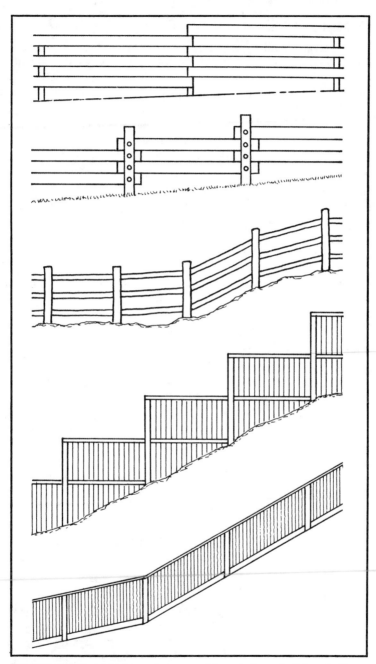

Fig. 5-18. Hillsides can present some problems in fence building, but here are five good ways to solve them.

boards fastened to the outside. Cleat boards over the boards add strength (Fig. 5-18).

Attaching Woven Wire Fencing

Metal fences can be erected on almost any kind of post. Often, for barbed wire and some types of woven and welded stock wire, wood posts are used. In other cases, metal posts are utilized. Because metal posts are of lesser diameter than wood ones, it is usually essential to provide some sort of bracing if the fence is of a heavy duty type. Sears offers an X-shaped anchor for its chain link fences that allows you to simply drive the post, with no digging required. The special driver that Sears will rent to you takes a lot of work out of that job. Figure 5-19 shows one way of bracing corner and end metal posts using concrete.

Once the posts are set, woven wire fencing is attached in sections that run from one anchor post to the next. As before, anchor posts include gate, end, corner and braced line posts. If concrete is used to brace posts, wait until it sets before keeping on with the job—about three days. Start about 2 feet ahead of your first anchor post and unroll enough of the fence to reach the second line post. Stand the fence roll on end. Remove two stay wires from the end of the roll (near the anchor post), getting a long enough piece to wrap around the post and come back on itself for a splice. Make the splices after wrapping the wire around the post, using the splicing tool shown in Fig. 5-20A and making sure the splice wraps around the wire at least five times. Also see Figs. 5-20B and 5-20C.

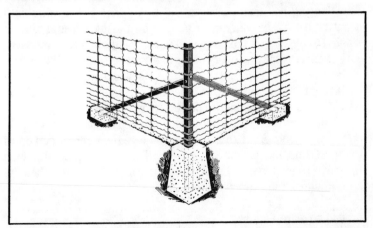

Fig. 5-19. Concrete used to brace steel end or corner posts will be set in this manner.

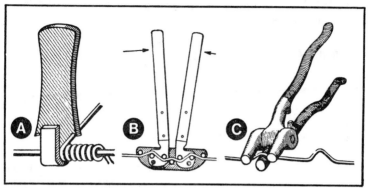

Fig. 5-20. Fencing tools. (A). Splicer. (B). A double crimper for taking up slack. (C). A single crimping tool for the same purpose.

Unroll the wire right to the next anchor post, and move back so that you are now some 6 to 8 feet beyond the second line post. Set a braced dummy post here (Fig. 5-21). Make sure the fence is solidly stapled or wired to the first anchor post now. Move along propping the fence against the line posts, using stakes to hold it in place if necessary.

Attach the stretcher unit shown in Fig. 5-21. Stretcher units have one or two jacks, and a single jack is used for fences under 32 inches in height. A double jack is needed for higher fences. Attach the stretcher unit so that wires are divided equally above and below or between the jacks. Apply slow tension to the fence, and keep a check on the wire mesh to make sure it is not caught on anything. Wire fencing has tension curves in it. You keep stretching until about a third of these curves are gone. Now the line wires are firmly stapled to the anchor post or wired in place if metal posts are used. This is done by cutting every other line wire, starting in the middle of the post, and wrapping it around the post and then making a splice back onto the line wire. Return to the middle and do the rest of the wires in the same manner. The top wire is done last.

Now move to the first line post and fasten the fencing there. Go on to the second. After that it's time to move the stretcher on down the fence line and repeat the process. The difference in attaching the wires to the line posts is that the top wire is done first, then the bottom wire, and then every other wire until all are fastened.

Grounding Wire Fences

Wire fences need to be grounded for lightning protection (of livestock, not the fence). According to the U.S. Department of

Agriculture, lightning striking a wire fence can travel as far as two miles along an ungrounded fence. Steel fence posts provide a ground for the fence only when they are in contact with moist soil, but for greatest safety extra long rods are needed. Figure 5-22 shows the correct method of grounding a metal fence with wood posts, and this method also works for metal posts. Use a ½-inch steel rod and drive it at least 5 feet into the ground, letting it stick up 2 or 3 inches above the top of the post. Fences should be grounded about every 150 feet this way. Grounding doesn't totally remove the danger of electrocution, but it greatly lessens it.

Fastening Barbed Wire

With all metal fencing, heavy leather gloves are recommended as are sturdy boots and tough clothing. These items are essential when working with barbed wire. Always stand on the opposite side of the fence from the stretcher, too, so if the wire breaks it will coil and snap away from you. Barbed wire usually doesn't have tension curves, so it is pulled up taut in most installations other than the barbed wire suspension fence (Figs. 5-23 and 5-24).

The steps are similar to those used for woven wire fencing, with the difference being that the wire goes up in single strands. The lowest wire is attached first, to the anchor post, and wrapped around and spliced on itself. It is then stapled (most all barbed wire fences use wood posts). The splice here will be four turns. If the anchor post is for a gate, the barbs must be removed to prevent

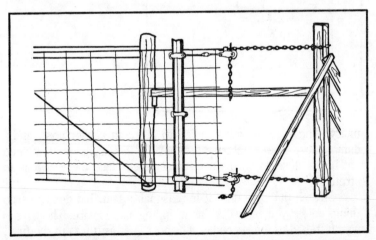

Fig. 5-21. A fence stretcher for wire mesh fence, showing a braced dummy post at the right.

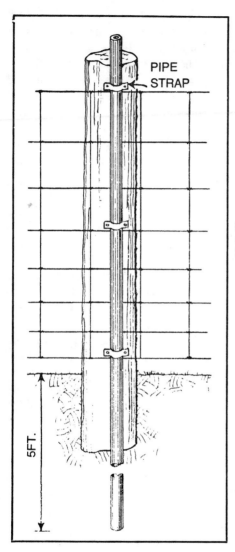

PIPE
STRAP

5FT.

Fig. 5-22. Grounding in this manner is essential for wire fences.

injury. The wire is now unrolled to the next anchor post, and a dummy post is set up about 8 feet beyond that anchor post. It need not be as close as with woven wire fence since there is less to stretch.

Stretch the wire until it is reasonably taut, but don't overdo things unless you enjoy ducking whipping barbed wire. The wire is now fastened to the second anchor post, just as it was to the first. Now the trip is made back down the line posts, and the wire is

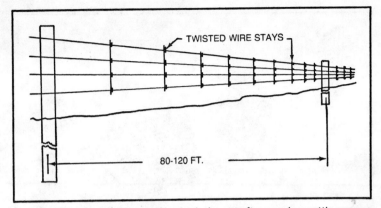

Fig. 5-23. Suspension fences of barbed wire are often used on cattle ranges.

stapled to the posts. Repeat with each strand of barbed wire and the job is done.

GATES

Many types of *gates* are to be found for fences. Usually, chain link fencing will have gates already constructed so that you merely need to hang them properly. Board fences may require you to build gates or to buy metal gates already constructed. These are available at farm and hardware stores and by mail order.

While the formed metal channels offer fair strength, they also seem to have a lot of sharp edges. Both Sears and Montgomery Ward offer hardware kits that ease the construction job of gates for board fencing, while most all post and rail fencing manufacturers make gates in the style of their fences. If none of this will do, write to the closest state-supported university in your state or check with

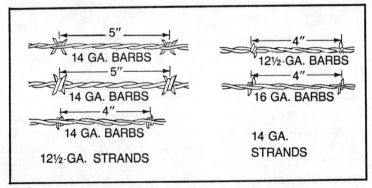

Fig. 5-24. Some common types of barbed wire.

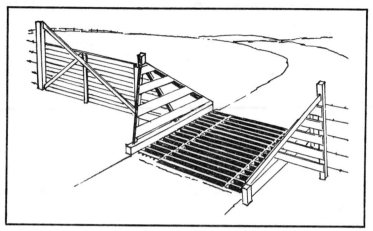

Fig. 5-25. Cattle guards are handy.

your county extension agent for specific gate plans to cover just about any possible contingency.

Cattle Guards

Cattle guards offer a simple way to keep cattle in a field while not forcing a vehicle driver to get out and open gates every time a field is changed (Fig. 5-25). Such a guard must never be used when horses are pastured in the same field, as a horse will attempt to cross and may break a leg or otherwise injure itself.

Cattle guards can be made of many materials, but should be constructed with vehicular use in mind. Planks set on edge can be used, but they should be at least nominal 2 by 10s in size. Pipes are a possiblity; they should be at least 2 and preferably 3 inches in outside diameter. Cattle guards should be at least 8 feet wide to accept vehicles, and the individual members should be no further than 3 inches apart. The pit underneath is 1 foot to 1½ feet deep and should be treated with old crankcase oil occasionally to keep weed growth down and prevent mosquitoes from breeding in any puddles that form.

Floodgates

A floodgate restrains livestock when your fence must cross a stream or gully. There are several ways to build these, and Fig. 5-26 shows one. End posts must be exceptionally strong and set far enough back from the stream banks so that they are not uprooted by erosion. The material hanging from the cable can be set solidly. If your area is subject to heavy rains, I would suggest using some type

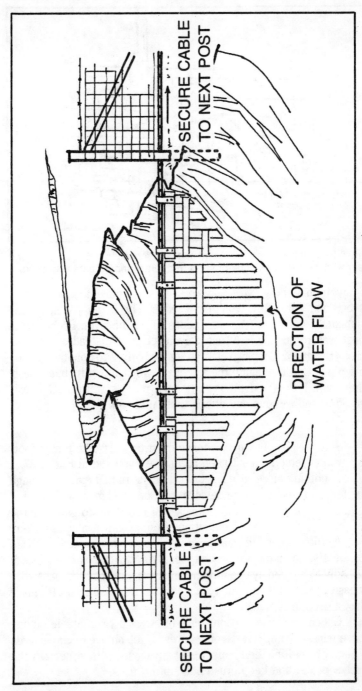

SECURE CABLE TO NEXT POST

SECURE CABLE TO NEXT POST

DIRECTION OF WATER FLOW

Fig. 5-26. When fences cross streams and ravines or gullies, floodgates will often be needed.

Fig. 5-27. A stile is useful on wood and wire fences to cut down wear and tear where a fence must be crossed frequently and no gate is needed.

of pivots that allow extra heavy resistance to move the gate so that debris and water flow won't rip it loose. The closer you space the slats in the floodgate, the more necessary is the pivot. Probably the best way to do this is to use galvanized pipe instead of cable to cross the gully. Then use clamps set fairly tight to hold the gate to the pipe. Cattle nudging the floodgate won't move it, but a flash flood will. If the floodgate is stationary, make sure to check every so often and remove any debris that builds up.

STILES

Stiles are handy where a fence must be crossed often on foot, but where vehicles don't need access. As you see in Fig. 5-27, a stile is nothing more than a set of rough stairs that extend through the fence. These crossing devices can't be used in areas fenced to hold sheep or goats as the animals will use them to get over the fence. In such cases, the installation of a small gate is much smarter.

A stile is a good idea in any area where a fence must be crossed frequently, no matter the fence type. Even sturdy board fences can be badly weakened by people constantly crawling over them. A stile is easier to erect than some types of gates. It is also easier to repair than a section of fence.

Check for loose sections once or twice a year. Staple on any loose wires. Crimp any sagging wires. Keep an eye on any boards that may be pulling their nails. Reset any loose posts, especially the anchor posts, and keep brush buildup down.

Chapter 6
Decks

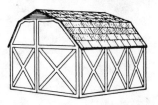

Last summer I added a deck to my home, and it has proven to be a great addition. I've slept there on summer nights and partied there on summer, fall and spring evenings. Plus, I've used it as a working base for remodeling the rest of the house. With a large sliding door, it makes an ideal outdoor workshop on good days and keeps the inside of the house free of sawdust and scrap lumber. A hose cleans it. It is made of Kopper's Wolmanized lumber, and is somewhat over-engineered since there is at least some chance that in the future the house might be expanded to cover the 12 by 24-foot size deck.

There's really nothing difficult about deck building, so it is ideal for the home handyman. A lot of nail driving through lumber is required. A bit of thought is needed when laying out the stairs, if any are to be installed. Commercially built decks are fairly expensive and pretty flimsy in my opinon. Yet even such a deck will add a couple of thousand dollars or more to the value of a house. How much will a properly constructed deck add? It's hard to say, but certainly much more than the cost in materials and labor (Fig. 6-1).

The deck that I built will be heavily used over the years and is actually quite a bit heavier and more expensive than most need be. Where I used 2 by 12s, most decks can do quite well with one or even two sizes smaller for the same span. I used doubled 2 by 12 beams, but most decks will be more than sturdy enough with doubled 2 bys 6s. My construction required over 35 pounds of galvanized nails; other will need possibly 20 pounds. The same holds true for the Teco anchors used; a size smaller is a bit cheaper. Figure 6-2 shows some deck building tools.

PLANNING

Planning a deck starts with lumber selection and brings us right back to pressure-treated lumber as the most economical and

Fig. 6-1. Plot factors can be readily adapted to provide unusual decks. This one is not at all suitable for homes with children.

durable way to go in most areas of the country. It's probably a good idea to take a look at the details of pressure treating wood. A good bit of your deck's endurance will be determined by the proper treatment methods, with the proper chemicals attaining the correct retention rate. Economy is aided, too, by not buying materials meant for in ground use for above ground use. Forget brush-on and soak type treatments if you really want the wood to last (Fig. 6-3).

Pressure Treatment Materials

Waterborne preservatives are pressure injected into the wood, and these types are the most commonly used around residences.

The chemicals used can be dangerous, but once into the wood they won't leach out and are quite safe. Still, it's advisable not to burn pressure-treated wood. This form of pressure treatment provides the wood with a characteristic light green cast which, over time, bleaches to a rather nice grayish color.

Creosote is also utilized in pressure treatment, for use in and on the ground. Such wood treatment makes the material suitable for use as poles, posts, landscaping ties and pilings in fresh water.

Pentachlorophends are dissolved in a petroleum carrier before being injected into the wood. The wood is suitable for the same uses as is creosote-treated wood.

Waterborne preservatives such as those developed by Koppers Company back in 1933 are suitable for almost everything you or I will ever need to do if we pay attention to the retention level. The greater the retention level, the greater the protection of the wood. Retention is measured in pounds per cubic foot of wood. Using Kopper's CCA chemicals, wood with a .25 pound per cubic foot retention are suitable for above ground use; .40 pounds per cubic foot make the wood suitable for use on and in the ground. If you're working in water, look for a .60 pound per cubic foot retention rate (fresh water).

Properly chosen pressure-treated wood, according to Gerald Wellner of Koppers Company, has an expected lifetime of no less

Fig. 6-2. A few of the essential tools for deck building.

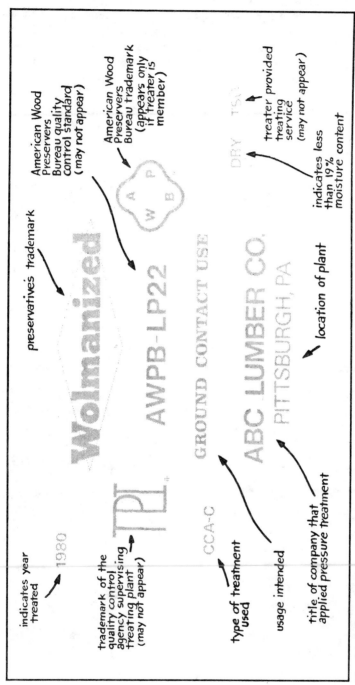

Fig. 6-3. Information often stamped on good pressure-treated lumber.

than 25 years. It is also resistant to termites as well as rot and decay.

Deck Use

Now that wood selection is done, it is time to consider other things such as the anticipated use of the deck. Will you use it only for sun bathing, or will it also be the site of parties and family cookouts? Each of these activities determines, at least in part, the size and design of the deck (Fig. 6-4). What sort of air currents are there around your home? Location of the deck can be affected by air currents. A gentle breeze is fine in the evening, especially on hot summer days. But a roaring wind in the middle of autumn is apt to reduce the fun of any party. How does the sun fall on your selected site? Do you wish a lot of sun, or do you prefer shade? Do you need privacy? Deck design can provide privacy with screens and other devices, but it is often simpler and cheaper to site the deck so that privacy is just about automatic. What sort of view is there? Mine

Fig. 6-4. Again, a tight plot with a pool can either provide a restriction or, in this case, a chance for the unusual.

overlooks a hillside plot of a couple of acres, with trees on three sides.

Safety considerations affects rail design. Infants need protection from falling, especially on decks on sloping lots, where the drop may be steep and high.

What sort of access do you want to the house? I had opened up the living/dining/kitchen area, removing one bedroom and moving the basement stairs. I removed the old porch door, installed a window, and set a 6-foot wide insulating glass sliding door in place of the old bedroom window. This gives immediate access to the present dining area, with the kitchen just a few feet to the left as you enter. Other people may well prefer access directly to a bedroom, kitchen or to the living room. Such access may be dictated by property lines and house location, not to mention local building codes. Don't forget to check on the need for building permits and code requirements for add-on construction, since most decks will be large enough to require permits.

Though decks first gained popularity as a way of adding outdoor living space on sloping lots, they are now becoming good party areas even on flat lots. It is possible to build a deck on-grade, just inches high, or off a second story. You can attach the deck to the house or build it as a freestanding unit, though I have yet to see one of these built very close to a house. Normally freestanding decks are built further from the house, with a walk leading to them.

While planning your deck site, check on the locations of any underground water and sewage lines. It would be embarrassing to crack through a sewer line or water main. In many areas electric and telephone lines are also buried, so it is a good idea to check on these.

Your code check will be more intense once the final decisions concerning size, shape and location of your deck are made. In many areas before permits can be issued, a decent drawing of the planned structure is needed. Some areas limit construction time to as little as 90 days. If you don't feel you can complete a large deck in that time, see what other arrangements can be made.

DESIGN

Deck *design* starts where house design does—with the footings. From there you move to posts, beams, joists, decking and railings. For a unit such as a deck which is primarily entertainment-oriented, attractiveness of appearance will also be important (Fig. 6-5).

Fig. 6-5. An example of the many designs possible for decks.

If the deck is to be a raised one, the location and size of the posts and beams will affect appearance almost as much as the decking and railing. Lower decks will not be so affected. I would suggest you take a look at a variety of decks in your locale, as well as looking at the plans here and in various magazines. I had certain railing ideas for my deck, but a ride in the rain with my brother changed them radically. He took me to look at a newly built house a few miles away. I didn't care for the stomach tossing qualities of a commercially built deck, but the railing design was easily adaptable to a sturdier form and much better than my original idea, though more wood and work was needed.

Railing choices are wide, but for strength it is wise to keep 2 by 4 railing uprights on 16-inch centers. However, a capping 2 by 6, with a 2 by 4 run under the cap as we did, can open this up to 3 feet or more if you wish. Four by fours capped by a 2 by 6 offer even greater spacing chances, up to about 8 feet (Fig. 6-6).

Posts, Joists and Beams

Decking is chosen for several qualities, including strength. Appearance is one. My preference is for 2 by 6s, but other people

113

Fig. 6-6. Check alignment before placing anchor bolts in railing and finish nailing the top rail to uprights.

prefer 2 by 4s. Some like 2 by 4s laid on their sides. Anything wider than 2 by 6s should be avoided for practical reasons. The wider the board, the more likely it is to warp. The pattern of the decking is pretty much up to you, and some ideas are presented in Fig. 6-7. Figure 6-8 shows a simple deck and should be studied to get an overall idea of the type of layout required. Once the overall size of the deck is determined, you'll need to figure the lumber sizes for the

Table 6-1. Minimum Beam Sizes.

LENGTH OF SPAN (FT.)	SPACING BETWEEN BEAMS (FT.)						
	4	5	6	7	8	9	10
6	4 × 6	4 × 6	4 × 6	4 × 8	4 × 8	4 × 8	4 × 10
7	4 × 8	4 × 8	4 × 8	4 × 8	4 × 8	4 × 10	4 × 10
8	4 × 8	4 × 8	4 × 8	4 × 10	4 × 10	4 × 10	4 × 12
9	4 × 8	4 × 8	4 × 10	4 × 10	4 × 10	4 × 12	*
10	4 × 8	4 × 10	4 × 10	4 × 12	4 × 12	*	*
11	4 × 10	4 × 10	4 × 12	4 × 12	*	*	*
12	4 × 10	4 × 12	4 × 12	4 × 12	*	*	*

* Beams larger than 4 × 12 recommended. Consult a designer for appropriate sizes.

various portions of the deck. We'll assume a deck size of 8 feet (span from the house) by 14 feet long, only a bit above ground level. Minimum beam sizes are shown in Table 6-1. We find, with 7-foot spacing of the posts, that the minimum beam size is a nominal 4 by 10. That means you can use two 2 by 10s, one on each side of the post.

Going to Table 6-2 to determine needed post size, we first find the load area (by multiplying the post spacing by the beam spacing). Our result is 36. The needed posts are 4 by 4s. The needed joist size, assuming we wish to use 32-inch centers, would be 2 by 8 (Table 6-3). From there, we look at Table 6-4 and see that 2 by 4s laid flat will do just fine with 32 inch on-center spacing, though 2 by 6s will be less springy.

All figures in the tables indicate a quality wood, with at least 1200 pounds per square inch bending stress rating, and a live load for the deck of 40 pounds per square inch. If the wood available doesn't meet this standard, or your anticipated load is to be higher, go at least one size larger for beams, joists and posts. Or close up

Table 6-2. Minimum Post Sizes.

HEIGHT (FT.)	LOAD AREA (SQ. FT.) = BEAM SPACING × POST SPACING				
	48	72	96	120	144
UP TO 6	4 × 4	4 × 4	6 × 6	6 × 6	6 × 6
UP TO 9	6 × 6	6 × 6	6 × 6	6 × 6	6 × 6

Vertical loads figured as concentric along post axis. No lateral loads considered.

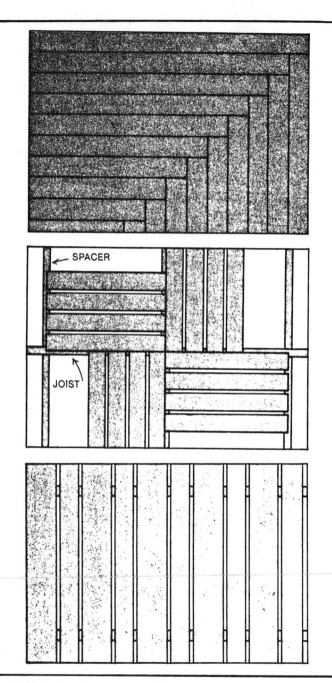

Fig. 6-7. Variations in decking layout are possible.

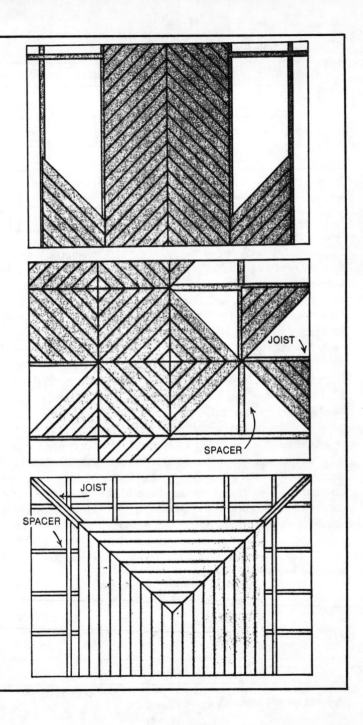

Table 6-3. Maximum Allowable Spans for Deck Joists.

JOIST SIZE (INCHES)	JOIST SPACING (INCHES)		
	16	24	32
2 × 6	9'-9"	7'-11"	6'-2"
2 × 8	12'-10"	10'-6"	8'-1"
2 × 10	16'-5"	13'-4"	10'-4"

your on-center spacing of the joists. The beams will still need to be larger, as will the posts.

Once the deck size, type and style is determined, you need a materials estimate, though some deck plans include that. Still, any variation from a standard design will require changes, as will any deck you design yourself. Start with a simple sketch of your deck, drawn most easily on graph paper. Try for ¼ inch to the foot scale for ease of working. When designing the deck, you'll usually save money by sticking as close to standard lumber sizes as you possibly can. You'll also save work in measuring and cutting most of the time. Deck boards in 2 by 4 and 2 by 6 sizes are usually stocked in 8, 10, 12, 14 and 16-foot lengths. These are, in fact, standard lengths for most kinds of lumber. Building to these sizes, or multiples of them, saves lumber, money, and cutting and measuring time.

Decking is laid with a space of about ¼ inch between the boards for drainage, and this should be figured in. Get the dimensions and quantities for posts, beams, joists, decking and railings. Most of the time lumber is sold retail on a unit basis, that is, so much per 2 by 4, etc. You may be curious to find the cost per board foot. Table 6-5 shows how to determine the board feet in various pieces of lumber, according to nominal size and total length.

Table 6-4. Maximum Allowable Spans for Spaced Deck Boards.

MAXIMUM ALLOWABLE SPAN (INCHES)		
LAID FLAT		LAID ON EDGE
2 × 4	2 × 6	2 × 4
32	48	96

Though able to support greater spans, the maximum spans will result in undesirable deflection or springiness in a deck.

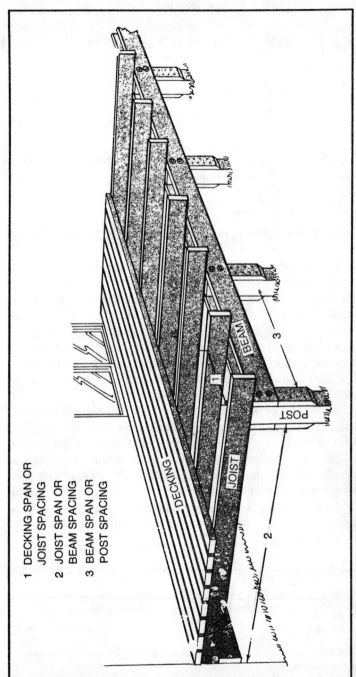

1 DECKING SPAN OR
  JOIST SPACING

2 JOIST SPAN OR
  BEAM SPACING

3 BEAM SPAN OR
  POST SPACING

DECKING

JOIST

BEAM

POST

Fig. 6-8. A simple deck design.

119

Table 6-5. Lumber Scale Board Feet Per Timber.

LENGTH OF TIMBER	8	10	12	14	16	18	20	22	24
1 × 4	2⅔	3⅓	4	4⅔	5⅓	—	—	—	—
1 × 6	4	5	6	7	8	—	—	—	—
1 × 8	5⅓	6⅔	8	9⅓	10⅔	—	—	—	—
2 × 4	5⅓	6⅔	8	9⅓	10⅔	12	13⅓	—	—
2 × 6	8	10	12	14	16	18	20	—	—
2 × 8	10⅔	13⅓	16	18⅔	21⅓	24	26⅔	—	—
2 × 10	13⅓	16⅔	20	23⅓	26⅔	30	33⅓	—	—
2 × 12	16	20	24	28	32	36	40	—	—
4 × 4	10⅔	13⅓	16	18⅔	21⅓	24	26⅔	—	—
4 × 6	16	20	24	28	32	36	40	—	—
6 × 6	24	30	36	42	48	54	60	66	72

Fasteners and Connectors

Your next materials estimation involves the *fasteners* and *connectors* needed for the job. Figure 6-9 shows the sizes of nails, while Fig. 6-10 shows various types of fasteners and connectors to make the job go faster and easier. These fasteners are essential to a good job. The joist connectors, post to pier connector, beam to post connector and rail to post connectors cut out much, if not all, the need for toenailing and other more difficult nailing techniques. Lag screws and carriage bolts secure the ledger board to the house sill when the deck is to be connected. Figure at least one lag screw every foot (actually two every 2 feet) when using a 2 by 8 ledger board.

Nails, as always, are to be three times the length of the board being nailed. Deformed, or ring shank, galvanized or aluminum nails must be used to prevent rust streaks. Pressure-treated wood will last longer than will untreated nails, which will eventually rust through and cause the deck to collapse. Some type of deformed shank nail is needed because of the effect of weather on the wood. The greater holding power is essential to a lasting job, unless you wish to spend hours each spring redriving nails.

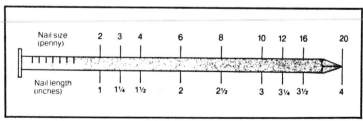

Fig. 6-9. Nail sizes and Lengths.

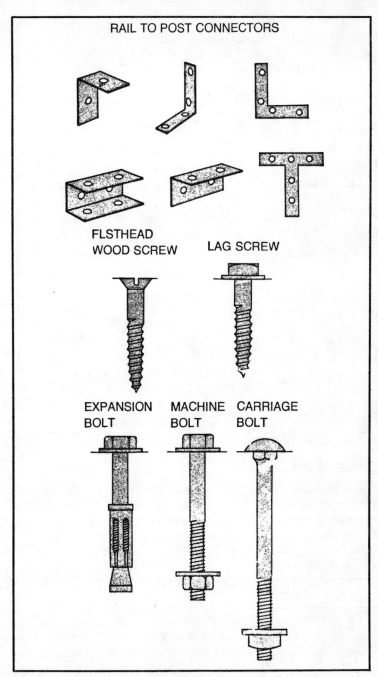

Fig. 6-10. Connectors for use with a deck.

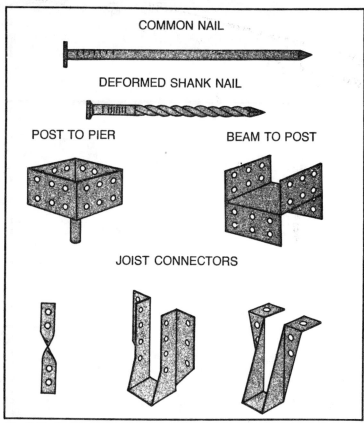

COMMON NAIL

DEFORMED SHANK NAIL

POST TO PIER

BEAM TO POST

JOIST CONNECTORS

Fig. 6-10. Cont.

CONSTRUCTION

The deck is laid out much as you would lay out any other form of rectangular construction. Use string and batter boards to mark out the area, keeping the string level and the corners square. To check that the strings are square once they are in place and level, use a felt tip marker to mark the string 3 feet in from one corner and 4 feet on the opposite, or adjacent, string. Then measure the diagonal, which will be 5 feet if all is square (Fig. 6-11).

Site Preparation

Site preparation for a deck is relatively simple compared to laying foundations and installing concrete footings. Take a spade and lift any sod from the area to be covered by your deck, going down about 3 inches and moving out some 2 feet beyond the deck's

122

proposed edges. Not much grass is going to grow under a shaded area such as a deck provides, so you might be able to use this sod on bare lawn spots. Now cover the area with a polyethylene sheet at least 4 mils thick. This will totally cut out weed growth. The sheet is now slit to accept the posts, and then covered with 2 to 3 inches of gravel, bark chips or other decorative material.

We assume you've checked for local frost depth, so the time has come to dig down for your footings. In no case should the postholes be less than 2 feet deep; in all cases they must go below frost line, plus 6 inches in which you place a 6 inch bed of gravel. Tamp the gravel firmly. You can also pour 3 inches or so of concrete instead of using gravel.

Now the posts are set, plumbed and braced. Plumb must be exact, and bracing should be sturdy enough to keep the posts from

Fig. 6-11. Layout and sod removal in preparation for constructing a deck.

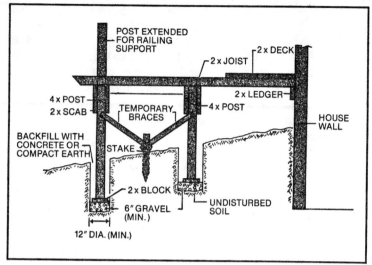

Fig. 6-12. Basic deck underpinning, showing bracing while waiting for concrete to set.

shifting while backfilling is done. You can backfill with the dirt removed already or you can fill the hole with concrete, which provides a sturdier overall unit. I used about 1½ feet of concrete in 2½-foot deep holes, filling in the last foot with dirt. I also used 4 by 6 posts on spans for which 4 by 4s are recommended (Figs. 6-12 and 6-13).

Securing Posts and Attaching Beams

Naturally, the posts must be aligned with one another quite carefully or you'll have trouble setting the beams in place. Use mason's string drawn tight, after setting the two corner posts, to align the inside posts.

Once the posts are secure, you can attach the beams. I find the easiest method of determining beam top height is to use mason's cord attached at the deck line on the house, drawn tight and level to the posts. Mark a line along the side of the post and transfer it to the front and back with a square to get the beam line. Actually, you must still subtract the actual dimension—not the nominal one—of the decking boards to get the correct beam height.

Much now depends on whether your posts will come up to support your deck railings. If not, from the beam line already drawn, subtract the actual dimension of the beam (width). Draw a line and, if the beams are to sit on top of the posts, make your cut. If the

124

beams are to be on each side of the post, cut at the marked height for the top of the beam. In cases where doubled beams are used to equal a single beam of their combined thickness, cutting at the top line provides you with a good nailing surface into the post. Figure 6-14 shows several methods of attaching beams to posts.

Installing Ledger Board

The *ledger board* now goes onto the house. This is held in place, usually on the sill of the house, with lag screws or carriage bolts. Where access inside the house is possible, the carriage bolts are used. Lag screws are used when no access is possible. Holes for lag screws are drilled slightly undersized and only about one-half to two-thirds the depth of the screw being used. Try a screw about twice as long as the ledger board is thick for greatest security. Then use joist hangers along the ledger board, though it is possible to toenail the joists directly to the ledger, or place the ledger under the joists and toenail into the house sheathing or studs. I prefer the hangers on the ledger.

When installing a ledger on lapped style siding, you'll find that it is necessary to either remove the bottom siding strip (often very

Fig. 6-13. Posts must be plumb.

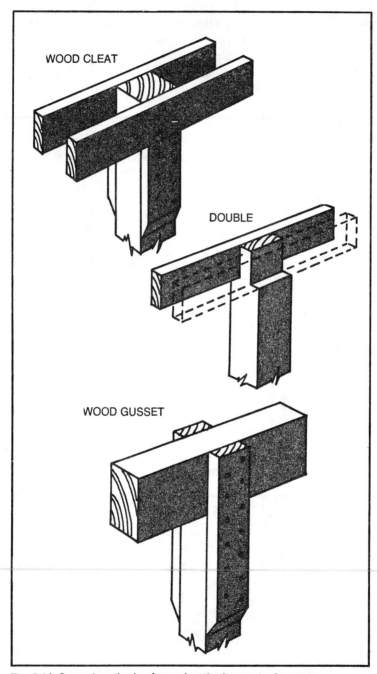

WOOD CLEAT

DOUBLE

WOOD GUSSET

Fig. 6-14. Several methods of securing the beams to the posts.

126

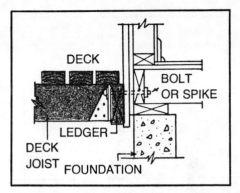

Fig. 6-15. Ledger board to house attachment.

difficult because it is the starter strip and nailing is seldom visible) or place another piece of the same style siding, upside down, on that first piece to give a level surface. If you do remove a piece of the siding, you'll have to slip flashing up under the next piece and run it down past the ledger board bottom to keep water from penetrating to the sill. Make sure the ledger board is level and stays that way during installation (Figs. 6-15 through 6-17).

The *joists* are now run from the beams to the house and nailed in place. Again, joist hangers can be used at the beam ends. My choice was different, since I wanted at least a 2-foot *cantilever* to serve as a place to store firewood at least partially out of the weather. I placed the beams so that the joists rested on them and then used Teco hangers to tie them to each other. This required the use of a *fascia* (facing) *board* on the ends of the joists, but is well worth the trouble. The fascia is a 2 by 12 so that the railings can be easily and solidly secured.

Installing Railings

Railings may be the next step, though it depends on just how you wish to install them. If the posts run up through the beams to

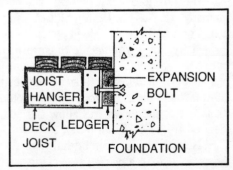

Fig. 6-16. Another method of attaching the ledger board.

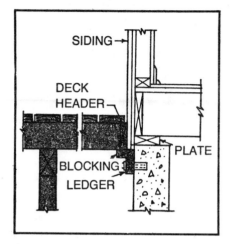

Fig. 6-17. A third method of attaching the ledger board to the house, using blocking to support a deck header.

serve as railing supports, now is as good a time as any to go ahead and install your railings. If you prefer to notch the railings to fit over the decking surface, then the deck boards are installed. Deck boards are installed using galvanized or aluminum nails, at least 12 penny in size. Again, a deformed or ring shank nail style is needed. Deck boards are spaced about ¼ inch apart. You can either make a spacer or simply drop a couple of nails between the boards being nailed. The spacer probably is the neatest and easiest way to go. Removing the nails stuck tight between nailed decking can be a pain (Fig. 6-18).

Decking is laid bark side up, as shown in Fig. 6-19. This helps to prevent cupping of the boards.

As you move along the deck, spacing can be adjusted so that there will be no huge gap at the end of the deck. It is a good idea to measure every three to four boards laid, after a third of the deck flooring is in place, so that adjustments don't have to be huge.

It is possible to buy different and decorative nail head patterns in many areas of the United States, and you may wish to consider using one of these for added appearance. This would probably be especially true if you decide to pattern your decking in something other than a simple joist-to-joist layout.

A deck installed with straight planking can be trimmed flush after installation. Any overhang left should not exceed 1½ inches. A flush cut ot the joists will allow the addition of a fascia board, if desired (Figs. 6-20 and 6-21).

Now the railing can be finished. As you can see, railing patterns are at least as varied as decking patterns. For this reason, it is

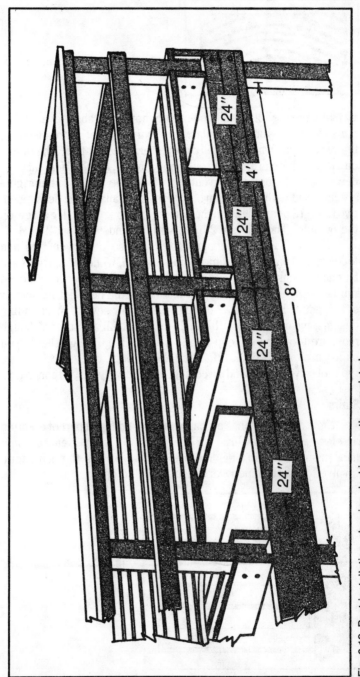

Fig. 6-18. Post installation, showing decking mostly completed.

129

BARK SIDE UP

Fig. 6-19. Lay decking bark side up.

virtually impossible to tell you excatly how to go about even laying out railings on your particular job. Your best bet is to sit down, figure what you want and then cut a pattern piece or two. The railing on the deck here is of 2 by 4s on 1-foot centers, notched for half their depth, with the bottom end cut at a 45 degree angle. The uprights are installed with three toenailed 12 penny nails on the deck sides, two more 12 penny nails on the outside, and one 3-inch lag screw on the outside. The cap is a 2 by 6, with an undersupport 2 by 4.

This rail is close enough together, so that even tots and toddlers will have a hard time falling through, while the uprights are far enough apart to prevent a child's head from being caught. And the rail is sturdy enough to support just about the heaviest leaning or sitting adult. Figures 6-22 through 6-25 show several more railing designs, including one with a built-in bench. All are sturdy, though most are too open to keep a child from falling through should you build an elevated deck. Figures 6-26 through 6-29 show various types of privacy and wind screening, all relatively easy to construct.

Stairs

Stair construction is probably the most difficult part of erecting an elevated deck, at least for the relatively inexperienced home handyman. It needn't be, for the hardest part is the arithmetic involved to determine riser height an tread depth.

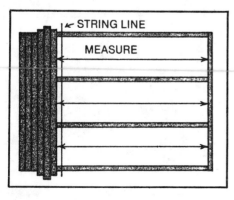

STRING LINE

MEASURE

Fig. 6-20. Decking can be laid along string lines.

Fig. 6-21. Snap a line, or lay a guide strip and cut decking off even.

Fig. 6-22. Post railings.

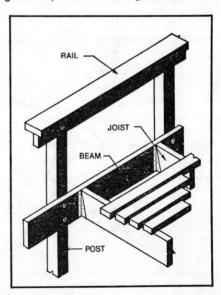

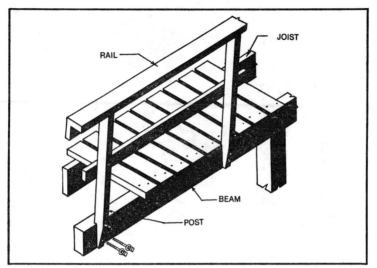

Fig. 6-23. Slant railings.

Stringers provide the support system for the steps and should be of 2 by 12 lumber, while treads may be of a single 2 by 10, a couple of 2 by 6s or three 2 by 4s. Much depends on the requirements for your particular set of stairs. Always remember that the wider the board, the more likely it is to cup or warp no matter how

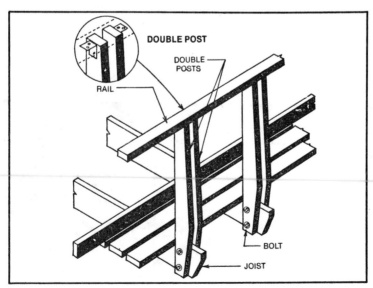

Fig. 6-24. Double post railing.

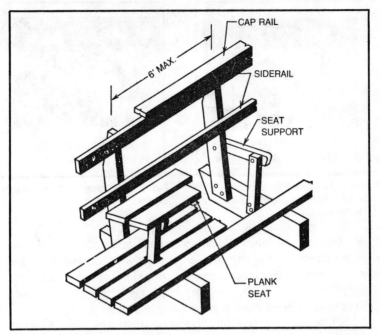

Fig. 6-25. Bench railings.

many nails are used. Figure 6-30 indicates some riser to tread proportions which give good footing and safety. It is best to always have tread depth and riser height as close to identical for each step as is possible. You'll note that the lower the riser height, the wider the tread.

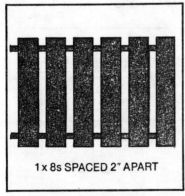

1 x 8s SPACED 2" APART

Fig. 6-26. A screen made by 1 by 8s spaced 2 inches apart.

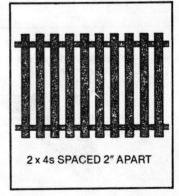

2 x 4s SPACED 2" APART

Fig. 6-27. Another nice screen is made of 2 by 4s spaced 2 inches apart.

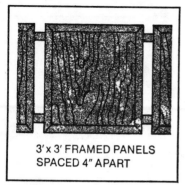

3' x 3' FRAMED PANELS
SPACED 4" APART

Fig. 6-28. Make a screen with 3 by
3-foot framed panels spaced 4
inches apart.

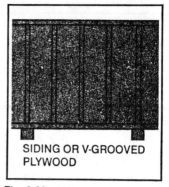

SIDING OR V-GROOVED
PLYWOOD

Fig. 6-29. Siding or V-grooved
plywood.

Stringers for cleated stairs, in which the treads are nailed to
cleats screwed to the stringers, are made of 2 by 10s only instead of
the 2 by 12s used for cutout stringers. Any stairway over 3 feet wide
should have a third stringer added in the center (Fig. 6-31). Stair-
ways over 6 feet wide will need a fourth stringer.

First, measure the total rise needed for your stairs. This is
done from the ground to the top of the decking, measuring vertically. Decide on riser height. Select a number that will divide
evenly into your total rise. Don't set things up so that the bottom, or
top, step rise is less or more than the rest. Then you can select a
tread depth, staying as close to a total of 18 inches as you can for the
two, riser and tread, added together. Remember, you'll want at
least a 1-inch overhang at the front of the tread (no more than 1½
inches, though). This overhang is *not* included in tread depth.

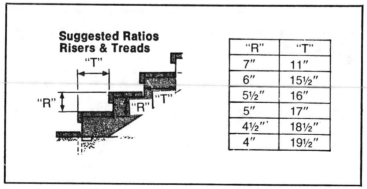

**Suggested Ratios
Risers & Treads**

"R"	"T"
7"	11"
6"	15½"
5½"	16"
5"	17"
4½"'	18½"
4"	19½"

Fig. 6-30. Suggested ratios of risers to treads.

134

For cutout stringers, use a framing square to mark the tread and rise for each step. Make sure on the last rise (for the bottom of the stairs) to cut away the distance used for tread material thickness. If the tread material is a nominal 2 by 4, or series of them, you will mark and cut off 1½ inches.

Coat the cuts with preservative. Nail the stringers to the deck, using either Teco anchors (simplest) or toenailing with 16 penny nails, using at least three on one side of each stringer and two on the inside.

Somewhat simpler to make, cleated stairs also need lighter material (by one size) for the stringers. But cleats are best screwed to the stringers, instead of being nailed as is too often done.

Fig. 6-31. Triple stringers are needed for stairs much over 32 inches wide.

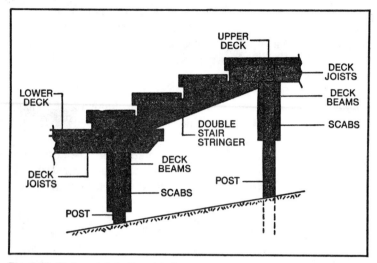

Fig. 6-32. Stair construction detail.

Measuring and marking is the same as for stairs using cutout stringers, but the only cuts needed are the top and bottom angles. There is less chance of destroying some fairly expensive material here, and the ease of adjustment is greater. Simply mark the stringers, screw the cleats in place, place the stringers and check. If all isn't well, you can remove one, two or more cleats and change their positions (Figs. 6-32 through 6-34).

BUILDING TIPS

Here are some deck building tips:

- Always nail thinner members to thicker members.
- Drive nails at a slight angle towards each other to increase holding power.
- When toenailing from opposite sides of a board, stagger the nails so that they won't strike each other.
- Use only annular or deformed shank nails and make sure they are rust resistant.
- To cut down on wood splitting, carefully tap the points of nails with the hammer to blunt them.
- When using lag screws, place a flat washer under the head.
- Wear gloves when handling wood. Even with planed wood, splinters can be a real pain.
- Don't burn pressure-treated wood. Dispose of it in an approved landfill, or bury it.

Fig. 6-33. Stair stringer layout.

FRAMING SQUARE
RISE
RUN

● Bevel the tops of uncapped railing uprights and posts extending above the deck at least 30 degrees. This helps water to run off instead of penetrating the wood (Fig. 6-35).

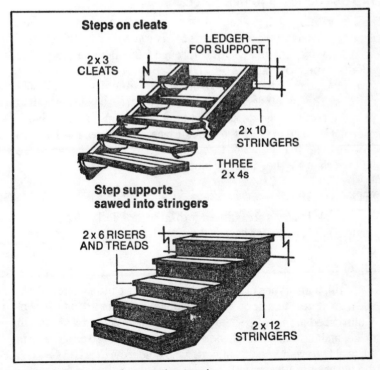

Steps on cleats

LEDGER FOR SUPPORT

2 x 3 CLEATS

2 x 10 STRINGERS

THREE 2 x 4s

Step supports sawed into stringers

2 x 6 RISERS AND TREADS

2 x 12 STRINGERS

Fig. 6-34. Two ways of supporting treads.

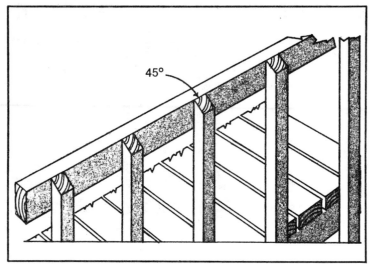

Fig. 6-35. Bevel railing taps to help promote water runoff.

TRELLIS-COVERED AND FIRE PIT DECKS

In addition to the general rectangular deck which is used as an example through this chapter, the *trellis-covered* deck in Fig. 6-26 and the *fire pit* deck in Figs. 6-37 through 6-39 offer good examples of other types. See Tables 6-6 and 6-7. Both are designed for near level plots, though the trellis-covered deck can be adapted to almost any plot. The fire pit deck should be built as a freestanding unit. It is placed directly on grade so that a large amount of concrete block need not be used to reach the deck surface. In addition, the 16 by 16 foot deck shown in Fig. 6-40 through 6-44 offers good adaptability to lots which are either level or sloped.

DECK FURNITURE

Simple benches are quite easy to build for a deck. Most people, though, seem to prefer lawn and garden furniture bought commercially.

Utility Table

Still, some types of furniture require only a little work and can add a great deal to a deck. As an example, consider a *utility table* of Wolmanized pressure-treated lumber. It takes about two hours to make, and the materials should cost no more than $20 or so. Size is adjustable to your needs, but here we'll hold to an overall 14½-inch top for ease of illustration.

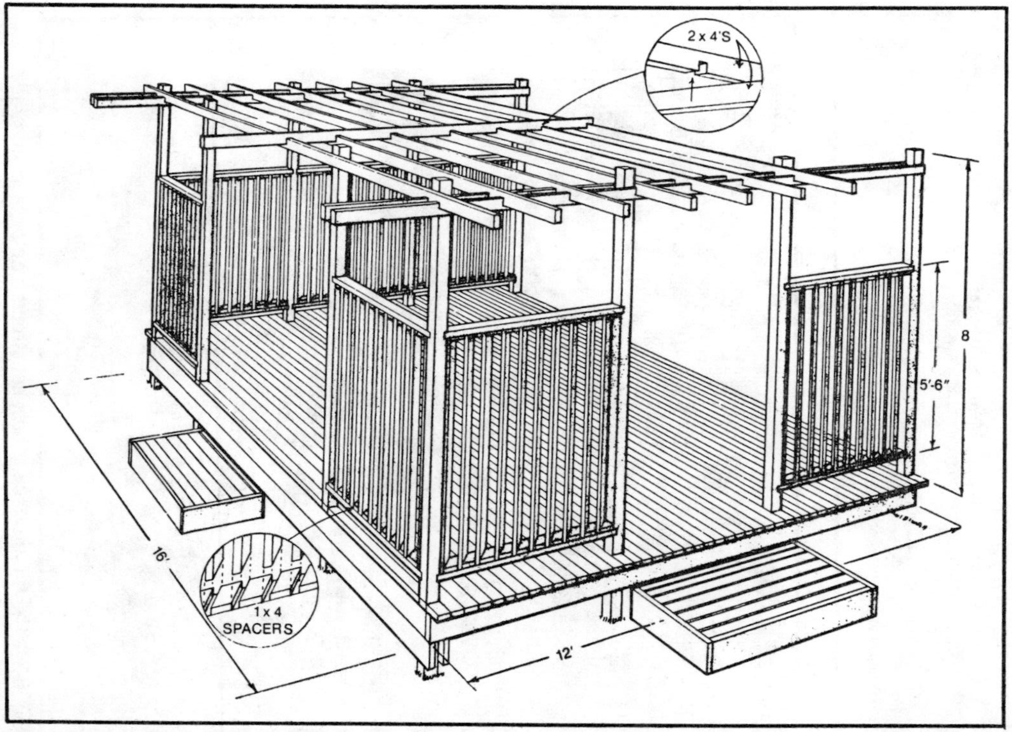

Fig. 6-36. This trellis-covered deck provides privacy on small plots.

139

Table 6-6. Materials List for the Trellis-Covered Deck.

1. Main support posts: 10 pieces required. Cut to fit terrain of your deck (approx.)
   10 pieces 4 × 4 × 10′

2. Materials for header boards: 3 pieces required 2 × 8 × 16′

3. Fascia boards:
   2 pieces required 2 × 8 × 12′

4. Joists:
   7 pieces required 2 × 8 × 12′

5. Decking:
   36 pieces required 2 × 4 × 16′

6. Trellis materials:
   Support beams: 5 pieces required 2 × 4 × 14′
   Trellis boards: 7 pieces required 2 × 4 × 18′

7. Privacy screen as per your requirements.

8. Nails and accessories as required.

Fig. 6-37. The fire pit deck is best constructed on or close to grade.

140

Table 6-7. Materials List for the Fire Pit Deck.

1. Deck:
    8 pieces required 4 × 4 × 10'
    27 pieces required 2 × 4 × 10'

2. Benches: (6'-8" long & 8'-8" long)
    5 pieces required 4 × 4 × 4'
    5 pieces required 2 × 4 × 7'
    5 pieces required 2 × 4 × 12"

3. Firepit:
    40 pieces required 4 × 4 × 16"
    1 bag mortar
    2 pieces required 2 × 6 × 6'

4. Nail and accessories as required

You'll need three 2 by 4s, 8 feet long; one 4 by 4, 25½ inches long; and a pound of 10 penny nails. A butt chisel is the only tool you may not already have that will be needed.

Cut five 11½-inch long pieces for the table top inside and two pieces 14½ inches long for the top's border. You'll now need four 18½-inch long pieces of 2 by 4, and 1½-inch square notches cut on the same side, in 6 inches from each end (these are for the table's feet). Four pieces of 2 by 4 are now cut into triangles: 4 inches by 90 degrees by 8 inches by 26 degrees by 8 15/16 inches by 64 degrees. It is best to use a miter box for these cuts, but they can be made freehand if you buy a good miter square. These triangles are top reinforcement gussets (Fig. 6-45).

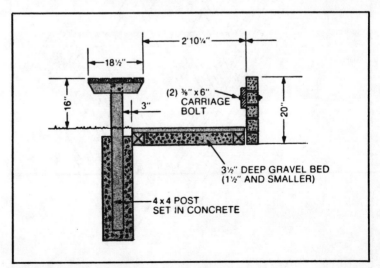

Fig. 6-38. Dimensions for the fire pit deck.

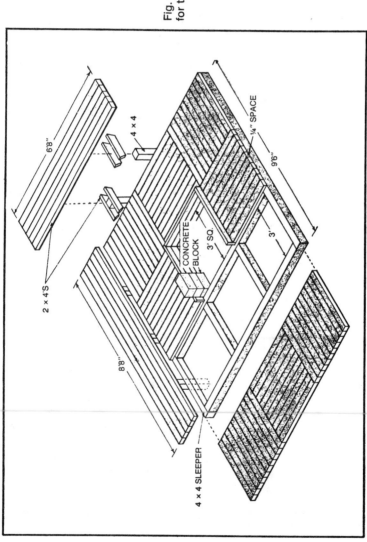

Fig. 6-39. Construction details for the fire pit deck.

When removing the material from the notches on the feet, the butt chisel is used. Keep the flat side of the chisel blade toward the notch seat.

Assemble the feet around the bottom of the post and insert one nail through the predrilled holes in the half-lap joint. Drive two nails, again through predrilled holes, through the center section of each foot and into the post.

To start the table top, predrill nail holes in both ends of the table top border. Nail the border together, using two of the table top interior pieces. Now take the remaining three table top interior pieces and nail them to the border, keeping ¼-inch spacing between the pieces. Make sure the table top is assembled so its surface is flat.

Next predrill two holes in each of the 24 degree and 64 degree ends of the gussets. Make sure the holes are drilled far enough back on the taper so that the nails won't stick through the table top. Set the table, top side down, on a flat and solid surface. Put the top of the

Fig. 6-40. This 16 by 16-foot deck is adaptable to many lot shapes.

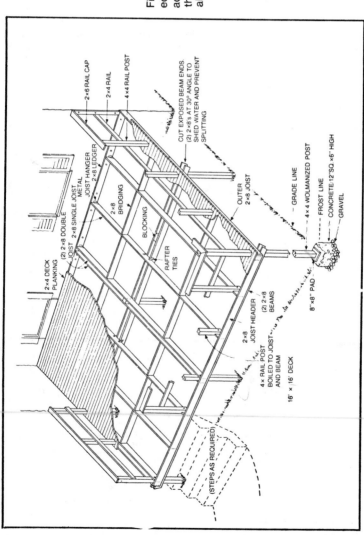

Fig. 6-41. The center and front edge beams are doubled for added strength, along with the center joist and two joists along the perimeter.

2×6 RAIL CAP

2×4 RAIL

4×4 RAIL POST

CUT EXPOSED BEAM ENDS
(2) 2×8's AT 30° ANGLE TO
SHED WATER AND PREVENT
SPLITTING

2×8 LEDGER

METAL
JOIST HANGER

(2) 2×8 DOUBLE
JOIST 2×8 SINGLE JOIST

2×8
BRIDGING

2×4 DECK
PLANKING

OUTER
2×8 JOIST

BLOCKING

— GRADE LINE

— 4 × 4 WOLMANIZED POST

FROST LINE

CONCRETE 12"SQ.×6" HIGH

GRAVEL

RAFTER
TIES

(2) 2×8
BEAMS

2×8
JOIST HEADER

8"×8" PAD

4 × RAIL POST
BOILED TO JOIST
AND BEAM

16' × 16' DECK

(STEPS AS REQUIRED)

post in the center of the table top, and arrange the gussets symmetrically. Get someone to help hold the post in place, and plumb, while you drive the nails through the predrilled holes into the bottom of the table top. Now, nail the gussets to the table top.

If you wish to increase the size of the top just a few inches, the legs can be left their present size. For increases much over 6 inches in top size, I would recommend also increasing the length of the legs for greater stability. Should the increase be on the order of twice the present size, it is best that you also make the gussets a few inches longer.

Simple Planter Box

A simple *planter box* can be an attractive addition to a corner of your deck, and is easily built in not much more than two hours. If you

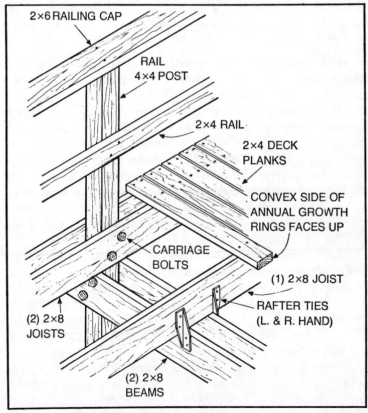

Fig. 6-42. This view shows how the doubled 2 by 8s are bolted to the 4 by 4 posts.

145

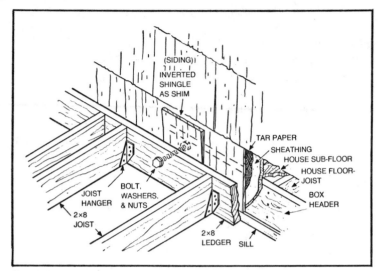

Fig. 6-43. Attaching the ledger plank to the house frame.

decide to use screws to build the box, the tool list will include a saw, hammer, carpenter's square, 6-foot rule, pencil, two bar clamps, power drill with a #10 combination bit, screwdriver and some sandpaper. Materials needed will be two pieces of Wolmanized pressure-treated wood 2 by 8 by 4 feet 3 inches long for the sides of the box; two pieces of 2 by 8, 7⅞ inches long for the ends; and a single piece of 2 by 6, 3 feet 9 inches long for the bottom. Also required is a piece of ½-inch wood dowel 9 inches long if screws are to be used. You'll need eight #10 by 3-inch long round head wood screws. If you decide to nail the planter box, get a pound of 16 penny galvanized nails. A single 14-foot length of 2 by 8 will supply all the material for the planter box.

Start by measuring and cutting the sides, ends and bottom, to length only for the bottom, from the long board. Now cut the angles, as shown in Fig. 6-46, for the sides and ends. Don't rip the bottom piece to width yet, as you'll get a more accurate fit by waiting until the sides and ends are assembled. If nails are used, simply nail the sides and ends together, rip the bottom piece to the correct width and nail it in place. See Fig. 6-47.

For a stronger box, clamp the pieces together in their final position and drill pilot holes for the screws. Now, remove the clamps and drill screw clearance holes in the sides only. Use the pilot holes as centers. Now you can counterbore for the dowels, using a ½-inch diameter bit and going to a depth of ¾ inch.

146

The ends and sides are now screwed together. The dowels, cut about 1/16 inch too long, are coated with a waterproof glue and inserted.

Once the glue has set, you can sand the dowels flush with the sides, rip the bottom to size and insert it. Drill a couple of drainage holes of small diameter in the bottom, and you're ready to start using the planter box.

If you decide to finish the planter box, use a lead free stain or paint. Do not finish the interior with anything else, such as varnish, as it could harm your plants.

Square Planter Box

Perhaps all the angles on the planter box keep you from building it. Well, there is a somewhat simpler way to provide for larger plants on or around your deck or in your home. The *square planter box* requires two 2 by 4s 14 feet long and two more 10 feet long, as well as one 2 by 8 by 10 feet. Get about three pounds of 8 penny galvanized nails, Now gather together a saw, square, hammer, wrench and a bit of sandpaper (Figs. 6-48 and 6-49).

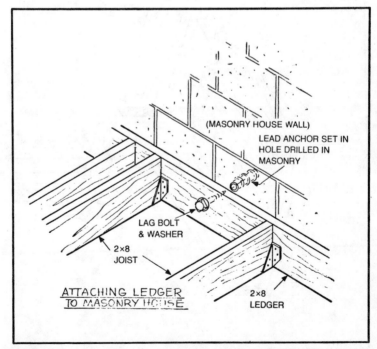

Fig. 6-44. Attaching the ledger ot the masonry house wall.

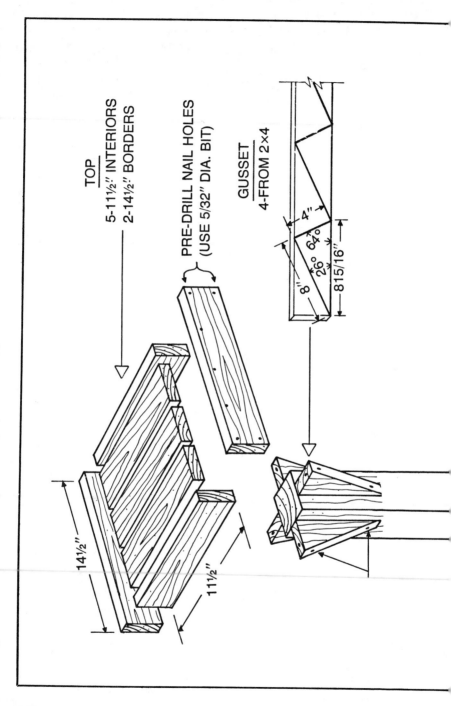

TOP
5-11½" INTERIORS
2-14½" BORDERS

PRE-DRILL NAIL HOLES
(USE 5/32" DIA. BIT)

GUSSET
4-FROM 2×4

4"

64°

26°

8"

8 15/16"

14½"

11½"

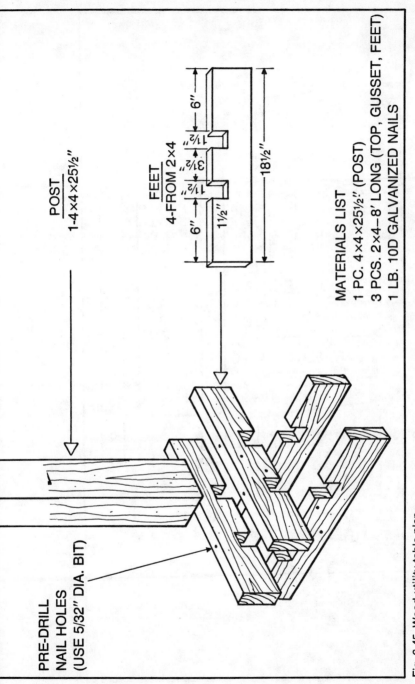

POST
1-4×4×25½"

FEET
4-FROM 2×4

6" | 1½" | 3½" | 1½" | 6"

18½"

1½"

PRE-DRILL
NAIL HOLES
(USE 5/32" DIA. BIT)

MATERIALS LIST
1 PC. 4×4×25½" (POST)
3 PCS. 2×4–8' LONG (TOP, GUSSET, FEET)
1 LB. 10D GALVANIZED NAILS

Fig. 6-45. Wood utility table plan.

149

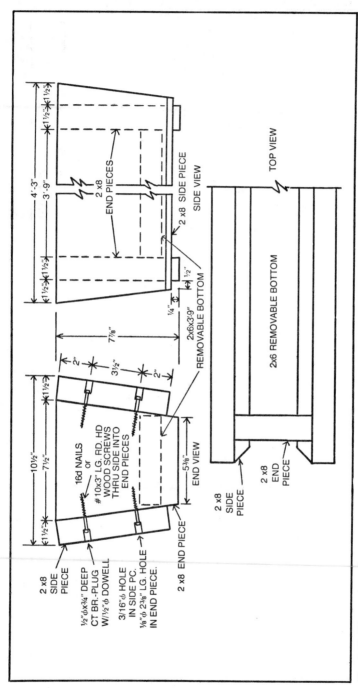

Fig. 6-46. Planter box plan.

Start by cutting four pieces of 2 by 4 to 22-inch lengths. Nail them together to form a square base. The sides of this base will now be 23 ½ inches long. You are now ready to cut six more pieces of 2 by 4 to 23 ½ inches to form the floor for this portion of the base. The 2 by 4 slats are nailed parallel to each other right on across the frame, leaving ½-inch gaps between the slats. Make a wood spacer for accuracy in gapping.

To make the top half of the base, cut four pieces of the 2 by 8 to 22-inch lengths and nail them together. The section formed is 23 ½ inches on each side. It will be 7½ inches high. Trim is cut now, with 16 pieces of 2 by 4 cut to 9-inch lengths and nailed vertically to the 2 by 8 portion of the frame. Start by nailing one piece flush with one edge. Leave a 3¼-inch gap and nail the next piece. Now leave a 3⅛-inch space and nail on another, and nail one more with the same space. The final piece of trim for the side will have a 3¼-inch gap and will come flush with the edge of the frame. Do this around the rest of the sides, keeping the trim even with the edge of the base. At the top edge, the trim will extend 2½ inches beyond the 2 by 8 frame. Now you take other portion of the base of 2 by 4s, turn it upside down an insert it into the 2 by 8 section. It is nailed, horizontally, to each piece of trim and also down through the 2 by 4s into the 2 by 8 section.

The entire box is now flipped into its right side up mode, while you cut four pieces of 2 by 4 to 23½-inch lengths. Lay them flat and

Fig. 6-47. Your flowers will look even more beautiful in this planter box.

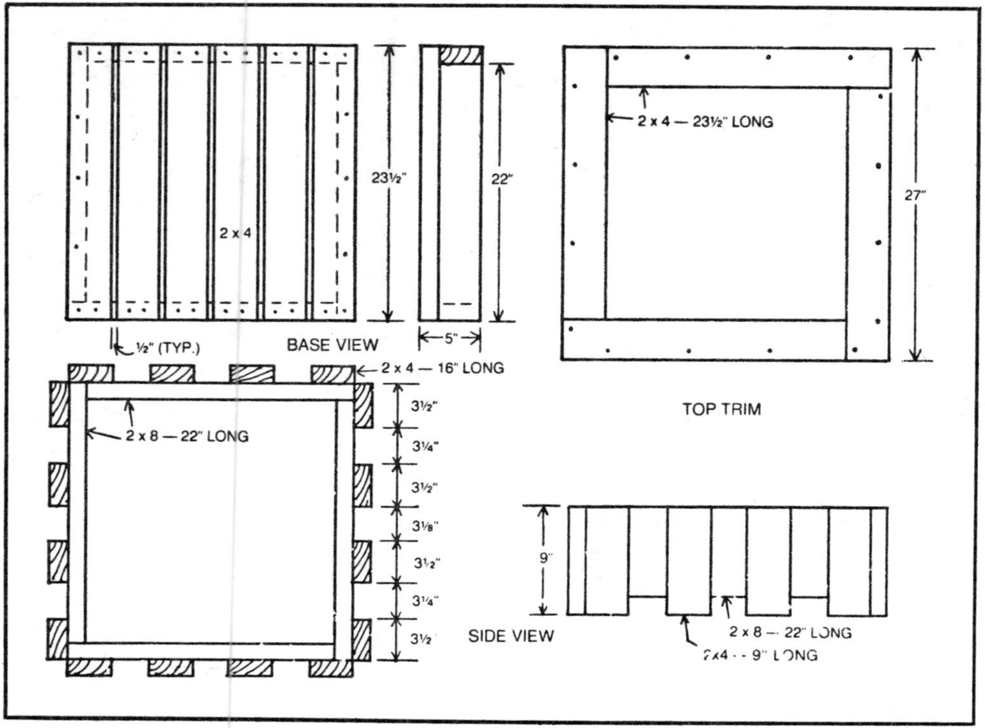

Fig. 6-48. Square planter box plan.

Fig. 6-49. The square planter box is a nifty looking item.

Fig. 6-50. Putting the last piece on the square planter box.

153

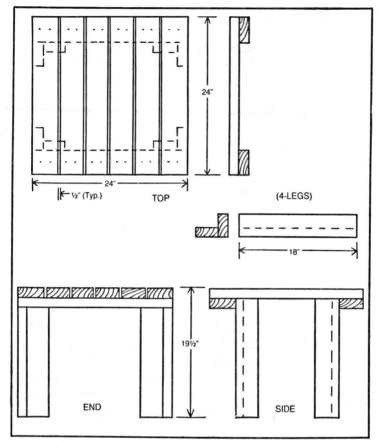

24"

24"

½" (Typ.)   TOP

(4-LEGS)

18"

19½"

END   SIDE

Fig. 6-51. Table-for-two plan.

nail them in place around the top edges of the planter. These are nailed both to the trim and the top of the base (Fig. 6-50).

Table For Two

A larger table is also handy at times. This one could prove useful just sitting in a backyard, as well as on a patio or deck. Select two 2 by 4s, 10 feet long; one 2 by 4, 14 feet long, and about a pound of (at least 56) 10 penny nails. You'll need only a saw, square, hammer and a rule, plus some sandpaper to smooth edges to do the job. See Fig. 6-51.

For the frame which serves as a base for mounting the top and legs, cut the two 10-foot long 2 by 4s into eight pieces each 23¾ inches long. Place four of these pieces on a level surface to make a square 23¾ inches on a side, making sure the corners are 90 degrees. It is best to predrill nail holes—use two nails at each corner.

The other four 23¾-inch pieces are placed inside the frame, allowing ½ inch or so between them for water run off. Nail them, with two 10 penny nails per end, to the frame underneath so that the table top is formed of six 2 by 4s.

The 14-foot board is now cut into eight 17¾-inch pieces so you can form the legs. Place one board on its side and lay a second flush along the top edge to form an "L" shape. Use four nails spaced evenly along the length. The result will be an L with one edge 6 inches long and the other 4 inches. Make four of these. Then place the table upside down and set the leg so the 6-inch side rests on the frame cross brace, with the inside of the L facing the center of the table. Fasten with four nails, two placed horizontally through the leg and into the underside of the frame and two vertically from the table top down into the leg along its outside edge. Once all four legs are in place, stand the table up, sand all the end surfaces and you're done (Fig. 6-52).

Bench

To make a deck *bench* of decent size requires nine pieces of 2 by 4 by 12 feet and two pieces of 2 by 12 by 12 feet, plus one 8-foot

Fig. 6-52. You can relax when the table is done.

155

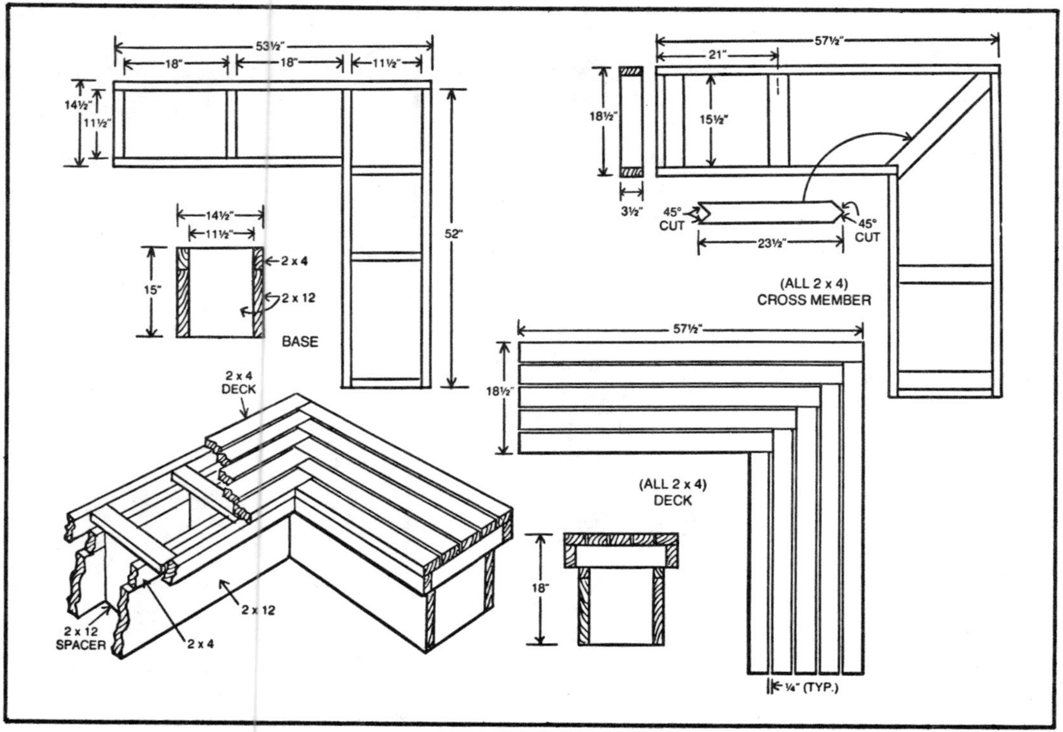

Fig. 6-53. Deck bench plan.

long 2 by 12. You'll also need about three pounds of 10 penny galvanized nails. Again, only simple tools are required: saw, square, rule or tape measure, hammer and sandpaper. See Fig. 6-53.

Start with the bottom half of the frame and cut one piece of 2 by 12 to 52 inches and another to 53½ inches. Nail together to form an L. Each side will measure 53½ inches if it is nailed correctly. Now cut two pieces of 2 by 12 to 15-inch lengths and nail one to each end of the longer boards, perpendicular to them and facing the inside of the L. Next, cut and insert a 52-inch section of 2 by 12 parallel to the other 52-inch piece. Nail where it meets the 53½-inch board. A 39-inch piece of 2 by 12 is now cut and nailed to form the shorter side of the L. You will now cut three 15-inch long spacers from the 2 by 12 material, and place one in the center of the shorter section of the shell. Its center should be 20¼ inches in from the outside edge of the endplate. In the longer section, the first spacer is placed 19 inches inside the outside edge of the endplate. The last is placed so its center is 38¼ inches in from that same end.

Fig. 6-54. Assembling the top frame of the deck bench.

Fig. 6-55. Putting the last pieces in place.

Now cut four pieces of 2 by 4 to lengths of 53½ inches, 37½ inches and two of 52 inches. Place them on the side boards with matching lengths to form support ribbing for the top, making sure the tops of the ribbing and the spacers are flush.

To begin the top half of the bench, cut six pieces of 2 by 4 to the following lengths: 57½ inches, 56½ inches, 40½ inches, 39 inches, and two of 15½ inches. Nail the two longest together to form an L, just as was done for the bottom part of the bench. Nail the 39-inch and 40-inch pieces together to form another L. Use the 2 by 4 by 15½-inch pieces to fasten the two Ls together. If the outside dimensions are correct, the end width will be 18 inches. Each long side will measure 57½ inches and the shorter sides will be 39 inches (Fig. 6-54).

For additional strength, cut five support beams. The first four need to be 2 by 4 by 15½ inches placed across the top of the shell—one at each end of the shell and the next two in so that they measure 21 inches to their centers from the edges of their respective endplates. All braces are placed flush with the surface of the shell. The final brace is 2 by 4 by 23½ inches, with 45 degree angle cuts at each end.

To make the angle cuts, draw a line 3½ inches back along the 2 by 4 to make a 3½-inch square. With a ruler, connect the opposite corners, forming an X within the square and cut out the bottom segment of the X. You now have the needed 45 degree angle. For the opposite end, measure 1¾ inches along the bottom edge and make a mark. From this point, draw a line to a point 1¾ inches up each side, forming a V. Cut along these two lines. Always cut just to the outside of any marked line—that is, to the scrap side. You'll have a pointed tip with a 45 degree angle. This is nailed into place on the frame.

Now that the frame is complete, you can take the top portion and fit it in place over the bottom section. It is nailed along the edges and through the cross bracing, since each brace should now be directly above a spacer.

To put a seat on it, cut 10 pieces of 2 by 4 to the following lengths: 57½ inches, two of 54 inches, two of 50¼ inches, two of

Fig. 6-56. Here's an interesting deck idea.

Fig. 6-57. This beautiful deck is a great addition to any home.

46½ inches, two of 42 inches, two of 52¾ inches, and one of 39 inches.

The 57½-inch piece is nailed along the top outside edge, flush with sides and ends. Nail one perpendicular to the board along the outer edge, using a 54-inch piece. Keep it parallel to the 57½ inch piece, leaving ¼ inch for water run off. Going back to the other side,

Table 6-8. Materials List for the Sun Deck.

1. Face plate:
   1 piece required 2 × 10 × 12'

2. Fascia:
   1 piece required 2 × 8 × 12'

3. Stringers:
   5 pieces required 2 × 8 × 12'

4. Diagonal braces:
   4 pieces required 2 × 8 × 3'3"

5. Beams:
   2 pieces required 2 × 6 × 14'

6. Post (length optional):
   3 pieces required 4 × 4'

7. Ledger strip:
   1 piece required 2 × 2 × 12'

8. Decking:
   39 pieces required 2 × 4 × 12'

9. Railing posts:
   10 pieces required 4 × 4 × 4'

10. Railing cap:
    3 pieces required 2 × 6 × 12'

11 Railing (center):
    3 pieces required 2 × 4 × 12'

12. Hot-dipped galvanized nails and hardware as required

(material for steps optional)

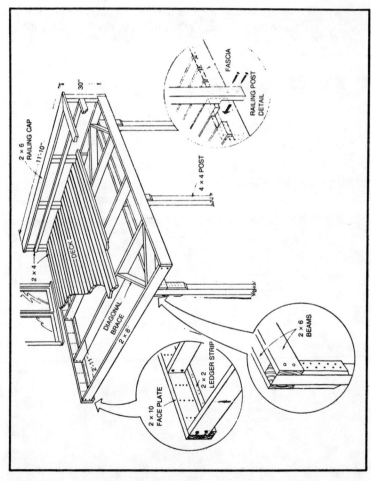

Fig. 6-58. Construction details for the 12 by 12-foot sun deck.

FASCIA

RAILING POST DETAIL

30"

2 × 6 RAILING CAP

11'-10"

4 × 4 POST

2 × 4

DECK

DIAGONAL BRACE

2 × 8

2'-11"

2 × 6 BEAMS

2 × 10 FACE PLATE

2 × 2 LEDGER STRIP

161

nail a 50¼-inch board in place. Continue on through the remaining planks until all 10 are in place. Go from longest to shortest and keep the planks flush with the edges of the frame (Fig. 6-55).

Using a wood spacer is best to make sure the ¼-inch interval is kept, as marking the interval for each board is very tedious. Sandpaper the ends' surfaces, and decide whether or not you wish to paint the bench, stain it or just let it weather. It won't rot if you leave it alone and will soon turn silver gray, freeing you from a possible yearly chore of staining or painting otherwise.

OTHER DECK IDEAS

And that's about it for decks and, at present, deck furniture. You may wish to use some later projects on decks, but their application is a bit more general. I've placed them in a different portion of the book for those people who already have, or don't wish to build, a deck (Figs. 6-56 through 6-58 and Table 6-8).

Chapter 7
Working With Brick

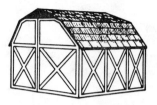

The use of brick for outdoor projects is nothing new. The material has been with us for thousands of years and may well be around forever. Brick is functional, attractive and possibly the most durable building material in existence when properly laid. Using brick sometimes frightens novices and often is passed up because of a perceived difficulty in doing a good job. Use care and you can build a good brick wall, patio, planter, barbecue or other projects.

BRICK SIZES

With some 10,000 styles of brick available, the variety of designs for any project are almost endless. Cost is naturally a factor, and it can be expensive if you hire a mason to do brickwork. Today's bricks range in cost from about a nickel each up to something like 20 cents depending on the size, style and distance the material must be transported. If you are fortunate enough to live close to a major brick manufacturer, the cost will be lower. There are major brick manufacturers in at least 44 states, with more in Canada.

For many years the following brick sizes were available—*Standard, Roman* and *Norman*. Brick is now available in sizes from a nominal 3 inches in thickness on up to as much as 1 foot. Heights range from 2 to 8 inches and lengths go up to 16 inches (Fig. 7-1).

While Tables 7-1 and 7-2 will show you some popular sizes available, not all manufacturers make these sizes. Many produce quite a few others. It is a good idea to check local availability when planning any job requiring brick. The sizes can relate to speed and ease of laying the brick and the total cost of the job under construction. In other words larger units go up more quickly and may require less mortar. Labor costs are cut, though the individual units used are more costly. Thinner units, especially in nonstructural applications, can mean better than fair savings in material costs on a job. Still, most outdoor construction is going to be structural in one

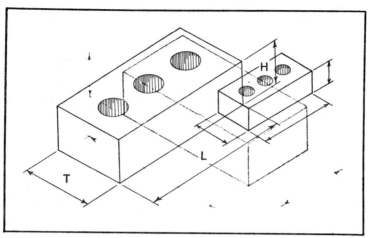

Fig. 7-1. Brick dimensions.

form or another, though a planter or barbecue may not be. Decorative brick walls and lawn edging are not structural, so the load bearing capacities of the brick are of little importance. Thinner units can easily and safely be used.

Nominal dimensions of brick equal the manufactured size plus the thickness of the mortar joint needed for that particular size and style of brick. In most cases, brick are laid with joints of either ⅜-inch or ½-inch thickness, so that the actual brick size will be that much less in length and height than the nominal dimension.

Table 7-1. Sizes of Non-Modular Brick.

UNIT DESIGNATION	MANUFACTURED SIZE, IN.		
	T	H	L
Three-inch	3	2⅝	9⅝
	3	2¾	9¾
Standard	3¾	2¼	8
Oversize	3¾	2¾	8

In recent years, the so-called "three-inch" brick has gained popularity in certain areas. The term "three-inch" designates its thickness or bed depth. The sizes shown in the table are the ones most commonly produced under the designation "Kingsize". Other sizes of 3-in. brick are also produced under such designations as "Big John", "Jumbo", "Scotsman" and "Spartan". Originally developed primarily for use as a veneer unit, it is also used to construct 8-in. cavity walls and 8-in. grouted walls.

The manufactured thickness of standard or oversize nonmodulator brick will vary from 3½ to 3¾ in. Therefore, if other than a running bond is desired, the designer should check with the manufacturer of the brick selected.

164

Standard modular brick, designed, to be laid three courses to 8 inches, will be 2¼ inches high, leaving a total of 1¼ inches of height for mortar joints. Figures 7-2 and 7-3 show some modular and nonmodular brick designs and dimensions. Figure 7-4 indicates the names for the various positions of brick as they are placed in a wall. Most brick will be laid as stretchers, with the longest of the face dimensions horizontal. The slightly shaded areas indicate the exposed portion of the brick.

BUYING BRICK

Brick is delivered to your dealer on pallets which will contain 500 bricks each (except for those with sizes varying greatly from standard). These pallets can normally be delivered directly to your building site and placed as you wish. I can stare at the three remaining pallets on my front lawn. Two of these are to be set in a walk, once I rebuild my front porch, and the third is for a project begun last year in the basement and awaiting completion. The basement style is open-cored, while the bricks to be laid as a walk are solid and oversize in thickness (width).

The delivery truck came equipped with an electric crane to lift the pallets and set them down, leaving the driver only the chore of accurately manipulating a boom with a set of buttons. Because I bought so many, delivery charges were minimal. It is also possible

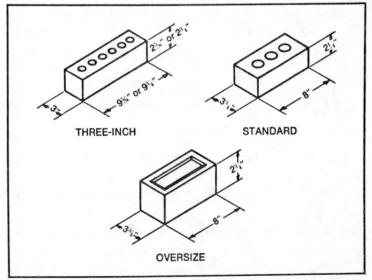

Fig. 7-2. Actual dimensions of non-modular brick.

Table 7-2. Sizes of Modular Brick.

UNIT DESIGNATION	NOMINAL DIMENSIONS, IN.			JOINT THICKNESS IN.	MANUFACTURED DIMENSIONS IN.			MODULAR COURSING IN.
	T	H	L		T	H	L	
Standard Modular	4	2⅔	8	⅜	3⅝	2¼	7⅝	3C = 8
				½	3½	2¼	7½	
Engineer	4	3 1/5	8	⅜	3⅝	2 13/16	7⅝	5C = 16
				½	3½	2 11/16	7½	
Economy 8 or Jumbo Closure	4	4	8	⅜	3⅝	3⅝	7⅝	1C = 4
				½	3½	3½	7½	
Double	4	5⅓	8	⅜	3⅝	4 15/16	7⅝	3C = 16
				½	3½	4 13/16	7½	
Roman	4	2	12	⅜	3⅝	1⅝	11⅝	2C = 4
				½	3½	1½	11½	
Norman	4	2⅔	12	⅜	3⅝	2¼	11⅝	3C = 8
				½	3½	2¼	11½	

Name								
Norwegian	4	3 1/5	12	⅜	3⅝	2 13/16	11⅝	5C = 16
				½	3½	2 11/16	11½	
Economy 12 or Jumbo Utility	4	4	12	⅜	3⅝	3⅝	11⅝	1C = 4
				½	3½	3½	11½	
Triple	4	5⅓	12	⅜	3⅝	4 15/16	11⅝	3C = 16
				½	3½	4 13/16	11½	
SCR brick	6	2⅔	12	⅜	5⅝	2¼	11⅝	3C = 8
				½	5½	2¼	11½	
6-in. Norweigian	6	3 1/5	12	⅜	5⅝	2 13/16	11⅝	5C = 16
				½	5½	2 11/16	11½	
6-in. Jumbo	6	4	12	⅜	5⅝	3⅝	11⅝	1C = 4
				½	5½	3½	11½	
8-in. Jumbo	8	4	12	⅜	7⅝	3⅝	11⅝	1C = 4
				½	7½	3½	11½	

to go to the dealer's, buy some bricks and bring them home yourself. If you do need more than a couple dozen, buying 500 brick lots is the cheapest way to go and provides a savings in energy for you. The lots need only be handled at the job site. Brick is far from the lightest building material known to man; carting bricks around more than is really necessary can take a lot of fun out of a project. Each standard brick weighs about 4 pounds, so a pallet will weight things down to the tune of one ton. A couple hundred bricks in the trunk of your car may lighten the front end enough to make driving dangerous.

ESTIMATING MATERIALS

Naturally you'll need to estimate your job's material needs before buying the brick. With most brick made on a standard of 8 inches by 3¾ inches by 2¼ inches, you can figure on each brick covering about 30 square inches laid flat on sand. A pallet of 500 will provide you with enough brick to construct a 10 by 10 foot patio if laid on sand. Using a mortar bed, with mortar joints of ½ inch, the patio size would increase to 10 by 12 feet.

For standard size brick when building a single brick wall, determine square footage of the wall and then simply multiply by seven. If walls are to be two bricks thick, multiply by 14. For lawn edgings, just multiply the length of the edging by 12, so you have the length in inches. Divide by the width of the brick you plan to use.

For jobs requiring mortar, every 500 brick will require 1⅓ sacks of Portland cement, 4 sacks of sand and ⅓ sack of hydrated lime. Each sack of cement contains 1 cubic foot. Mix just enough mortar to work about 40 bricks at a time—that amount is 1 shovelful of Portland cement, 3 shovelfuls of sand and ¼ shovelful of lime.

A few tools are needed for working with brick. Start with a mason's hammer and a brick chisel as these will be needed even if mortar isn't used. Then get a good tape measure. Add some mason's line if you're building a wall, as well as a good mason's level, mortar box and board, trowel, shovel and hoe for mixing the mortar.

LAYING BRICK

Working with brick is hard physically, but not otherwise complex for simple jobs if the rules are followed. As a start, one of those rules should be a check of local building codes if you're planning to build much more than a small barbecue or planter. First, this insures your project will be legal when completed. Second, if there are any special needs in your locale, such as extra deep or wide footings for walls, find out before you begin the project.

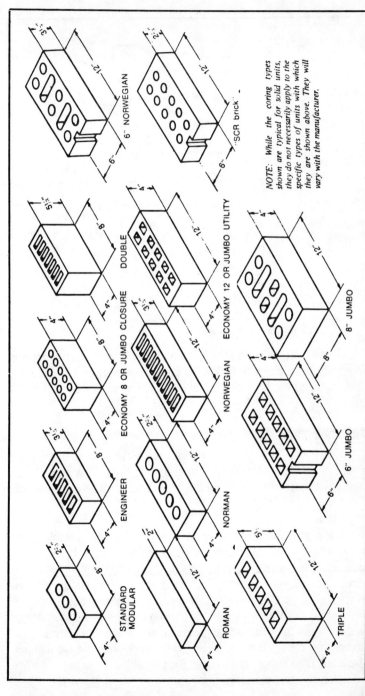

Fig. 7-3. Nominal dimensions of modular brick.

169

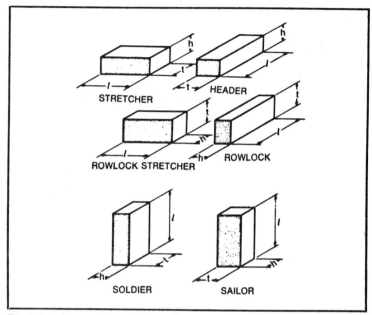

Fig. 7-4. Brick positions in the wall.

Have your materials ready. You should have at least 5 percent more brick on hand than the job estimate requires (to allow for breakage and math goofs) and about 25 percent more mortar mix materials.

Site Preparation

Start with site preparation and put in any needed footings. Footings, or other bases for brick laying, should have level surfaces. For mortared brick, these surfaces need not be dead smooth. For unmortared brickwork, a level and stable site is of very great importance (Fig. 7-5).

Now it's time to check whether or not you have a brick meant to be laid dry ar damp. Draw a circle on one of the brick, about quarter-sized, using a grease pencil. Put 20 drops of water in the circle, using a medicine dropper. If the water is still visible after 90 seconds, the brick should be laid dry. If the water is absorbed, the brick must must be dampened before laying or it will suck enough water from the mortar to prevent complete hydration. The strength of the bond will thus be reduced. The best way to dampen the brick is to simply hose down the pile about 15 minutes before you plan to start working.

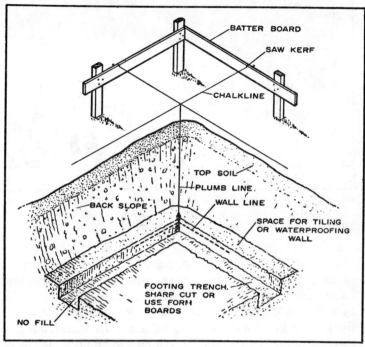

Fig. 7-5. Establishing corners for footings.

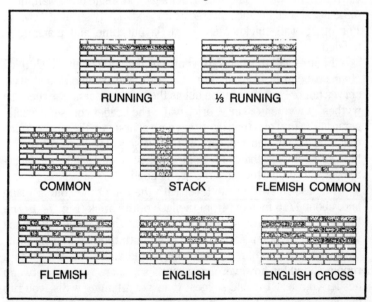

Fig. 7-6. Types of brick masonry bonds.

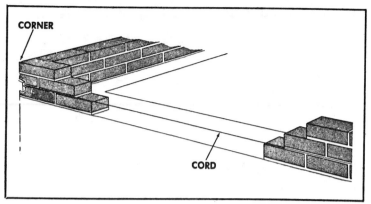

Fig. 7-7. Run the cord level with the top of the course being laid.

Running Bond and Common Bond Walls

For walls, the first course of brick is laid dry to check the number actually needed. For a running bond wall, each brick in succeeding courses will have its end over the center of the brick below. See Fig. 7-6 for other types of bonds. Use a stretched mason's cord to show where the outside edge of the wall must come. Brick are laid from the corners in, assuming your project has corners. Each corner should be built up several brick high before the rest of the wall is laid. For a running bond single brick wall, little difficulty should be encountered as it is extremely straightforward. For other wall thicknesses, more layout time and planning is needed.

Figures 7-7 through 7-10 show construction of an 8-inch thick common bond brick wall. For this wall, the common bond style serves not only as decoration but as the bonding agent between the wythes. A wythe is a single brick wall. The header courses, usually set in at the first and then every seventh course, tie the two sections together. As you can see, corners for this sort of wall are considerably more complex than for a single wythe running bond wall. Quarter and three-quarter closures are needed.

A mason's cord is run level with the top of the course being laid, about 1/16 inch out from the face of the brick (Fig. 7-7). The line doesn't need to be laid for the courses on the backing wall, since you can simply keep each course level with the one in front of it. Once the header course is laid, begin the wall. You can go to a running bond for however many courses you are running between header courses. I would suggest that, for retaining walls, you run header courses at every fifth or sixth course for added strength.

Since 8-inch walls are more likely to be used for such work than anything else these days, you'll seldom run out as many as seven courses before adding a header course. For extra heavy duty work, a foot thick common bond wall, combining one course of running bond with a header course, can be used. These are exceptionally sturdy, but time consuming and rather expensive to build since a lot of material is needed.

Trowel Work

On the footing, once the dry laid bricks are taken up, lay a bed of mortar about 1 inch thick. Furrow this bed with the trowel (Fig. 7-11). Now lay the first corner brick, shoving it deep into the mortar bed. Continue laying brick on both sides of the corner, going out at least five brick per side (unless the job is so small it won't be five bricks wide, of course). Lay only enough mortar to bed these bricks. Now lay a bed of mortar on top of the bricks and furrow it. Butter the end of the second brick, to go into place on the second course and shove it into place with a downward movement. This is going to produce what is known as a *shoved joint*, squeezing mortar out at the vertical and horizontal joints. The mortar bed on this second course should be only a bit thicker than the final joint thickness, probably ⅝

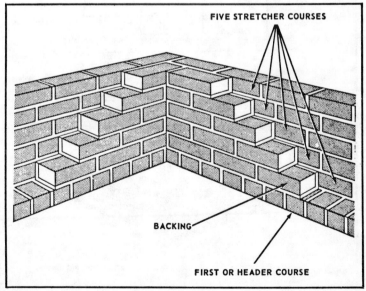

Fig. 7-8. Laying backing brick at the corner on an 8-inch common bond wall.

173

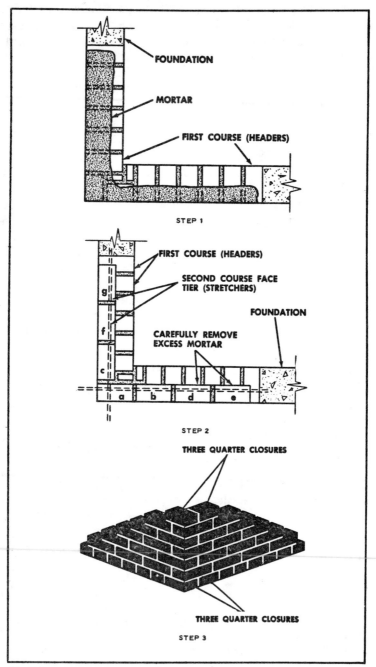

Fig. 7-9. The second course of a corner lead for an 8-inch wall.

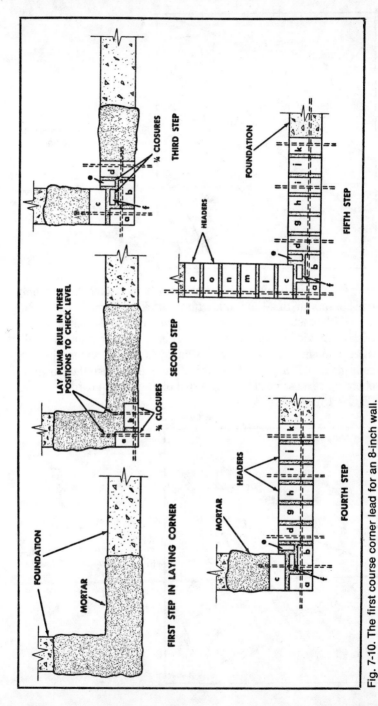

Fig. 7-10. The first course corner lead for an 8-inch wall.

175

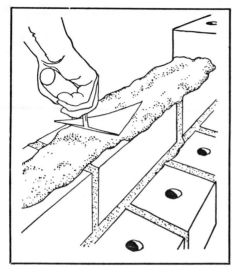

Fig. 7-11. Using a trowel to furrow the mortar bed.

inch for a ⅜-inch joint and ¾ inch for a ½-inch joint. Spread only enough mortar this time to lay three brick at a time (Fig. 7-12).

With practice you'll be able to shove a buttered brick into place without moving it once it is down. This provides a tight and well-filled joint and one that will help keep moisture out so that the joint remains tight. Once the three brick are in place, use the trowel to cut off any excess mortar and return it to the mortar board for re-use (Fig. 7-13).

Fig. 7-12. Shoving the brick in place.

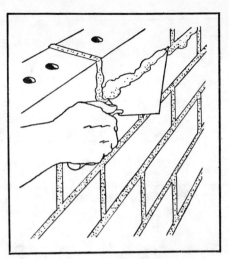

Fig. 7-13. Cutting off excess mortar.

Once all the brick are in place, use a jointing tool or the tip of the trowel to tool the joint for greatest weather resistance. This job is extremely important and not difficult. It is done when the mortar is thumbprint hard, meaning it will just take a thumbprint impression. Properly tooled joints will aid greatly in weatherproofing any brick wall (Fig. 7-14). The tooling packs the mortar tightly in the joint and prevents water entry. For exterior walls, a concave tooled joint is always best. Figure 7-15 shows several possible joint styles.

Often brick must be cut. A bolster or brick chisel is used in conjunction with a mason's hammer. The chisel is placed on the line to be cut and tapped. This scores the brick, and the chisel is then moved to a second surface of the brick to make another score line. Finally the chisel is placed on the score line, tilted to point in a bit,

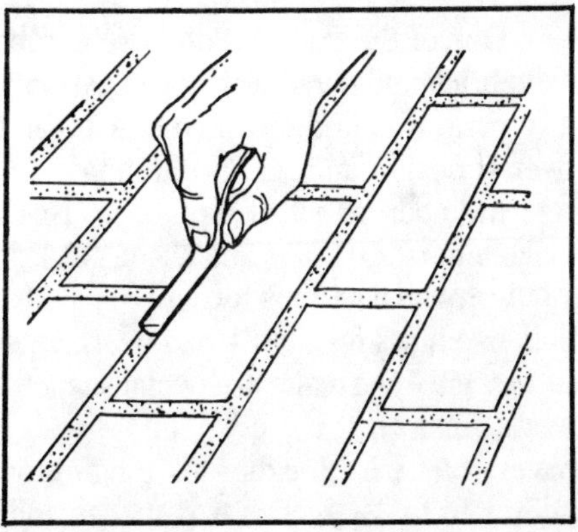

Fig. 7-14. Tooling the joints.

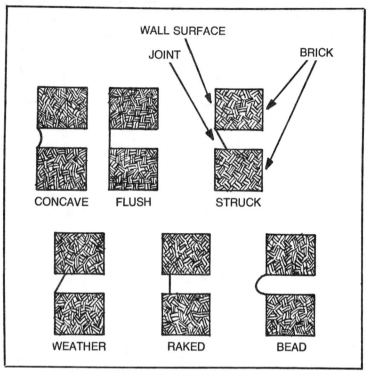

Fig. 7-15. Joint finishes.

and given a sharp blow. Usually this will result in a reasonably clean break. Any rough edges can be easily trimmed with the sharp edge of the hammer (Figs. 7-16 and 7-17).

BRICK PAVING

One of the earliest jobs I remember was helping a man clean used brick and lay a fairly large patio in New Rochelle, New York. The patio was laid on a bed of sand and contributed greatly to the overall design and feel of the home.

Member manufacturers of the Brick Institute of America today make almost 40 sizes and shapes of brick paving units (Fig. 7-18). Some specialty types for use in paving are also available, such as bullnose units for stair tread overhangs and so on. A few manufacturers will even make up special shapes on request, but such custom work tends to drive costs up. Bricks manufactured especially as pavers come in thicknesses ranging from 1½ to 2½ inches, with widths from 3⅜ inches to 4 inches and lengths from 7½ inches to

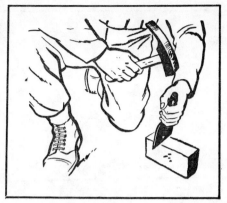

Fig. 7-16. Cutting brick with a bolster.

11¾ inches. Squares from 4 inches to 12 or even 16 inches on a side can also be found. Hexagonal bricks are made in 6, 8 and 12-inch sizes (Fig. 7-19).

Brick Grades

For exterior paving use, the grade of brick is of particular importance. The Brick Institute of America recommends SW grade brick for all exterior work, and nothing less should ever be used as paving brick. As a rule, dark brick is a hard burned brick and is considered more durable than light-colored brick. Today this rule is a bit outdated, as there are many SW grade light-colored bricks now

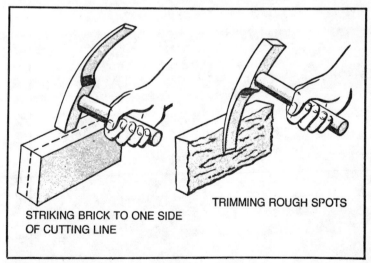

STRIKING BRICK TO ONE SIDE OF CUTTING LINE

TRIMMING ROUGH SPOTS

Fig. 7-17. Cutting brick with a hammer.

Fig. 7-18. Exterior brick pavers.

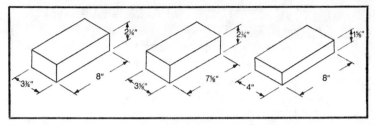

Fig. 7-19. Some popular brick pavers and their dimensions.

made. Paving brick is almost always uncored (solid), but cored SW brick can be used if placed on edge. This tends to eat up a lot of brick since coring is usually done on the widest portion of the brick. Brick used for mortarless paving works best if the units are twice as long as they are wide, but this is not completely essential if a ½ or ⅓ running bond pattern is used.

Drain the Bed Properly

There is no real secret to getting a good job without using mortar to hold the bricks in place. First, the bed must be firm and

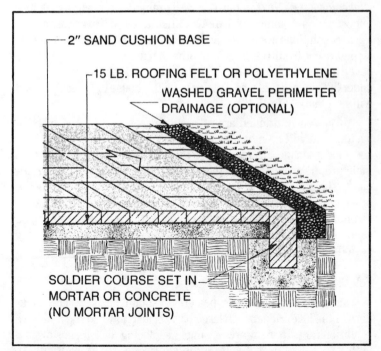

2″ SAND CUSHION BASE

15 LB. ROOFING FELT OR POLYETHYLENE

WASHED GRAVEL PERIMETER
DRAINAGE (OPTIONAL)

SOLDIER COURSE SET IN
MORTAR OR CONCRETE
(NO MORTAR JOINTS)

Fig. 7-20. Mortarless patio construction, using a sand base.

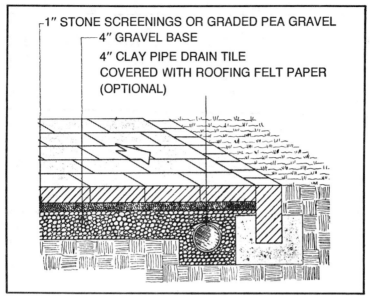

1" STONE SCREENINGS OR GRADED PEA GRAVEL
4" GRAVEL BASE
4" CLAY PIPE DRAIN TILE
COVERED WITH ROOFING FELT PAPER
(OPTIONAL)

Fig. 7-21. Using a pea gravel base.

well compacted. Next, it must be well drained. This last is the more important, and sometimes more difficult, of the two essentials. Various solutions are available. Brick paving installations should be sloped away from the house or other structure so that all of the paved area will drain away. If the patio is to be relatively small (say under 300 square feet), a fall away from the house of ¼ inch per foot will be plenty (Fig. 7-20).

In decently drained areas, a subsurface of 2 inches of compacted sand will be more than enough to provide both a good base and good drainage. In areas where there is a great deal of subsurface moisture, from a high water table or land that perks poorly (*percolation* is the ability of the soil to let water drain below the surface), other measures are needed to keep the moisture from moving up through the paving. Under such conditions, sand should never be used as a base because of its tendency to retain moisture for long periods of time (Fig. 7-21).

Pea Gravel Base

Instead, a layer of clean pea gravel should be used. The thickness of the layer depends in large part on just how bad the moisture conditions are. In many areas, simply replacing the 2 inches of sand with 2 inches of gravel will do the job. Cover the gravel with 15 pound roofing felt and lay the pavers on that, brushing sand into the joints to finish the job. For more extreme conditions, a 4-inch gravel base covered with 1 inch of pea gravel could be needed. For really extreme conditions, that base might have to be combined with clay drain tile piping set to carry a steady stream of water away from the patio and house. Usually such drain tiles are placed close to the edge of the patio and set in a deeper trench, with the top of the 4-inch drain tile extending about 1½ inches uu to the gravel base. At least 2 inches of gravel are underneath the drain tile. For exceptionally bad moisture conditions, you may need to use the setup shown in Fig. 7-22. The drain tiles are set under the low point of an inward sloping patio and into a trench at least a foot deep and filled with gravel.

Edging Brick

Once the moisture conditions are mastered, you can outline the area to be paved using mason's string and stakes. Edging brick is set first and may be placed in concrete or mortar to prevent shifting. Unmortared brick pavers do have a tendency to shift under use, but setting in concrete is not always essential to a good job. Cutting into the ground to leave the patio level with the surface will usually provide good support for longer bricks. Still, give some considera-

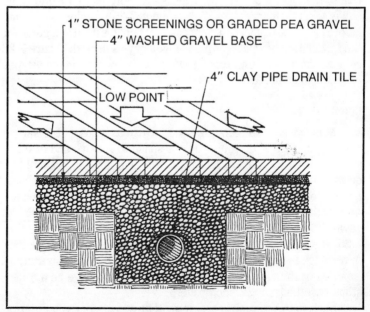

Fig. 7-22. A gravel base for severe drainage problems.

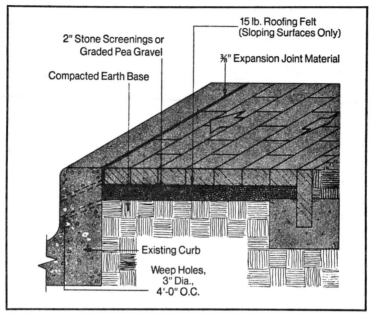

Fig. 7-23. Gravel base for a walk.

tion to the type of soil, as the edgers are more likely to shift if the soil is sandy or loamy (Figs. 7-23 through 7-25). Edging is installed before pavers are placed in most cases. Then it is used as a guide for the installation of whatever pattern you have selected. Figure 7-26 shows some possible pattern variations for standard and other types of rectangular brick.

Cored Brick

Mason's sand is recommended for use as a base. It is spread and tamped well after you've used a spade to remove sod or other material from the area to be paved. If gravel is used, it should be clean, preferably washed, and range in size from ¾ to 2½ inches. A leveling bed of small stone screenings or pea gravel is placed on top to provide at least a 1-inch thickness for the brick. The thicker the paving bricks used, the thinner the base usually needs to be. As Fig. 7-25 shows, a cored brick can be used on edge to give a 3½-inch paver thickness over just 2 inches of gravel. The Brick Institute of America has found this method to be just as effective as a 5-inch stone base under a 2¼-inch paver.

Such an installation is easier in other ways, too, as compaction doesn't require heavy equipment. The 2-inch base can be com-

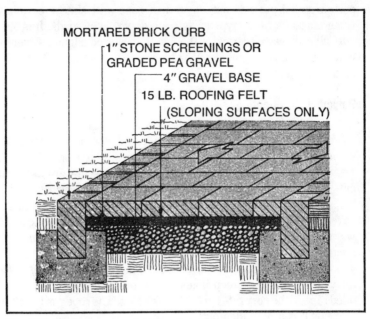

MORTARED BRICK CURB

1" STONE SCREENINGS OR
GRADED PEA GRAVEL

4" GRAVEL BASE

15 LB. ROOFING FELT
(SLOPING SURFACES ONLY)

Fig. 7-24. Mortarless construction for a driveway.

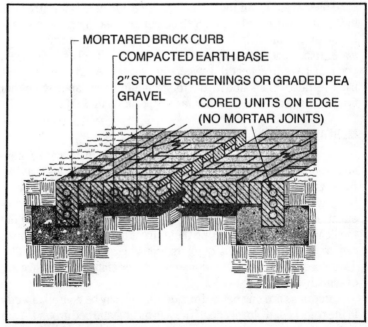

MORTARED BRICK CURB

COMPACTED EARTH BASE

2" STONE SCREENINGS OR GRADED PEA
GRAVEL

CORED UNITS ON EDGE
(NO MORTAR JOINTS)

Fig. 7-25. Using cored brick on edge for a drive.

pacted well by hand, but the 5-inch base will almost certainly require a roller of the type used to put down paved roads. It is best with all driveways to provide them with mortared edging, as vehicular weight and movement will be more likely to cause shifting if the edgers are just set in place.

Polyethylene Sheeting

You'll note in several of the drawings that a membrane of roofing felt is shown. The membrane could also be at least 4 mil polyethylene sheeting. It is used for several reasons. First, it makes the paving units easier to set in place accurately. Next, it cuts down on the rise of subsurface water into the bricks, which helps cut back on any tendency toward staining the bricks might have. If the patio, drive or walk is correctly sloped, any excess water getting trapped between the units and the membrane will eventually drain off or evaporate. If the slope is greater than normal you will want to, if the edging is set in concrete, supply *weep holes* for water drainage. Weep holes are made by inserting pieces of oiled rope in the concrete base for the edging. The rope can be left in place or it can be pulled out.

Place the brick pavers in your selected pattern, and then sweep sand over them to fill the joints left. If you want greater strength, sweep a mixture of 1 part Portland cement and 3 or 4 parts sand between the units. In both cases, the pavement is sprinkled with water to compact the sand between the bricks. If only sand is used, you may have to repeat the sweeping process several times as the sand settles. As an alternative to sprinkling, you can leave either type of installation to weather, as a fine rain will do the same job.

MORTARED PAVING

Laying brick paving with mortar joints tends to be more expensive and a lot more time consuming than the mortarless styles. First, the integrity of the entire unit depends on having an inflexible base, which won't be moved by frost heave. While any frost heave effect on mortarless brick paving is easily corrected by lifting out the heaved units and resetting them, frost heave in mortared units can crack entire sections of joints, causing all sorts of difficulty. Therefore, proper construction methods for slabs and footings are essential.

Footings are dug below frost depth and may be poured directly in the trenches, if cleanly dug. They go on undisturbed ground. Your slab will go in over at least a 2-inch gravel base, with drainage attended to as I described earlier. A membrane of polyethylene sheeting over the gravel, or even under it, is a good idea. Once the slab and footings have cured, the brick can be laid. You'll find that curing the slab takes three to seven days, with seven days of dampness preferred.

Slab thickness depends on paving use. Generally, for a driveway with brick being laid in a full mortar bed over the concrete, a 3 to 4-inch thick slab is sufficient. This assumes decent footings, a good base and a driveway not subjected to use by heavy trucks. Such vehicles as fuel oil trucks could cause mortar joints to crack. Once the slab is in, the bricks are laid in a full mortar bed at least 1 inch thick. Mortar joints can be either ⅜ or ½ inch thick. Use mortar

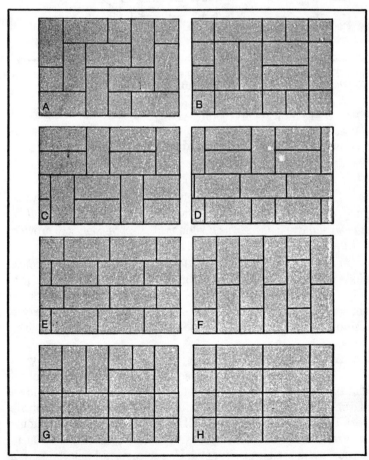

Fig. 7-26. Varied paving patterns for brick.

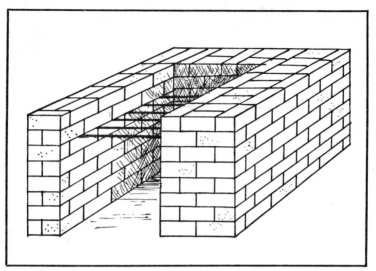

Fig. 7-27. Mortarless barbecues lend themselves to quick construction and easy design changes, both during and after building.

mixed of 1 part Portland cement, ¼ part hydrated lime and 3 parts of sand. Joints should be tooled carefully. It is probably best to compact the mortar well without making a concave joint.

MORTARLESS BARBECUE

The variety of projects possible with brick could really defy the imagination. Let's look at building *barbecues*, a *retaining wall*, a *mailbox stand* and a *screen wall*, along with *stepping stones* to provide a pathway where a full walk isn't needed or wanted.

For mortarless patios, a *mortarless barbecue* is a convenience since you won't have to dig a foundation and pour concrete for it (Fig. 7-27). The first job is to select the type of brick. Firebrick are not needed here. Next comes the selection of the grill racks, as the size of these may well determine just how you'll erect your barbecue.

About the only tool you'll need to work this barbecue is a good 2-foot level, though you might need a brick chisel and hammer if dimensions don't fit the full size of the bricks selected. Select a level and stable site for the barbecue—directly on the already laid patio is best. Besides the grill racks, you'll need about 250 brick, depending on size, to build this unit. Because the barbecue has no mortar to support it, the walls must be solid. Use at least one brick course as a base to keep hot coals from the patio. Lay out the base in your pattern, using at least a double brick construction allowance where the unit will be. Bricks are in an interlocking pattern, and the bricks to hold the grill racks are set in about 1 inch to provide support surfaces. Make sure these bricks are set in before laying the exterior brick to surround them. The barbecue is now ready for use.

SCREEN WALL

A *screen wall*, as shown in Fig. 7-28, can provide privacy to a porch, lawn are or patio. This wall will require 4.4 units of 3¾ inch by 2¼ inch by 8 inch brick per square foot in the screen areas, and

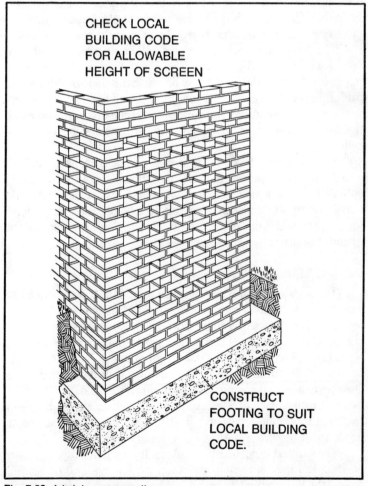

CHECK LOCAL
BUILDING CODE
FOR ALLOWABLE
HEIGHT OF SCREEN

CONSTRUCT
FOOTING TO SUIT
LOCAL BUILDING
CODE.

Fig. 7-28. A brick screen wall.

6.55 bricks per square foot in the solid areas. Such a screen wall is not allowed in a few areas. Be sure to check your local codes.

Dig the footing trench to the correct depth for your area, keeping the width to just what is needed for the footing so you won't have to build forms. Keep the bottom of the footing trench as level as you can, using a 4-foot level to maintain a check. Pour your footing and give it at least three days of dampness to cure properly.

Once the footing is cured, lay out the first three courses of brick dry. It is essential that this type of wall have its corners just about dead plumb, so a careful check must be maintained. Once the solid wall base is laid, you start the pierced pattern by cutting a brick to 6-inch length (Fig. 7-29). Next, cut a brick in half and lay it in place, centered on the point where two bricks come together to form a joint. Place the cut portion of the brick to the rear of the wall.

The next course is begun with a half brick, which is followed with a whole brick (Fig. 7-29). The whole brick will span to the middle of the cut brick in the row below. Stretcher, or full length, brick make up the rest of this course. The half brick pattern starts again in the next course. The patterns are repeated until you reach the selected height or the height allowed by your local building codes.

While building the screen wall, take as much care as possible to avoid putting pressure on the screen portion of the wall. Since it is only half as heavy as a solid wall, it is also less strong and could tumble or deform under pressure applied before the mortar sets. Joints are tooled in one of the concave patterns, again using care not to shift the brick.

MIALBOX STAND

This *mailbox stand* is cantilevered to hold the box and is a fairly complex project requiring a good footing and some use of reinforcing rods. Because some shifting would be allowable, the footing needs to be only 16 by 24 inches in perimeter and 8 inches deep. It should still reach below frost line to get greatest durability. You'll need about 60 cored bricks in 3¾ inch by 2¼ inch by 8 inch size. Quarter-inch pencil rod reinforcing is cut to the following lengths: two of 4 inches, two of 6 inches, four of 8 inches. You'll also need two lengths of reinforcing bar, ½ inch by 53 inches long. The mortar needed will require about 5.3 cubic feet of concrete.

The mailbox is not erected in place. It is built in an upside down position and then set into place when the footing is to be poured around the ½-inch reinforcing bars. Whatever work surface is used,

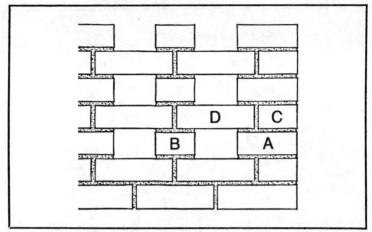

Fig. 7-29. The pattern formed by solid and cut units.

it must be level and stable. It should be as close as possible to the final site of the mailbox since the stand will have two legs, with each weighing about 130 pounds. Carting that much weight any further than is needed is no fun.

The first three bricks are placed, as shown in Fig. 7-30, using thin wood strips to hold them in place. The vertical mortar joints hold them together at the start. Now fill a core with mortar and insert one of the 4-inch lengths of pencil bar. Cut a brick in half and lay in place along the full length units. Lay two full units, and fill the cores to take the 6 and 8-inch bars. Fill only those cores where the reinforcing bars are used. The next course takes two full bricks; now insert the 8-inch bars. Cut another brick in half and lay it in place. Fill the cores and insert the ½-inch rod. If this rod moves too much, the mortar won't bond to it properly. Let it sit for at least three hours before coming back to the job. You can start the second leg and go to the ½-inch bar point, while waiting for the first leg to set around the bar.

From this point, spread mortar on the brick and carefully set each one over the ½-inch bar. Cores are filled completely each time a brick is laid. When you reach the correct number of courses to provide your needed height, prop the leg on all four sides so no shifting is possible. Let the leg stand for at least one week to insure a good cure and long lasting bonding.

Once the week has passed, pour the concrete footing. Bend the ½-inch bar to a 90 degree angle so that it will clear the bottom of the footing hole. Slip a piece of pipe over the bar and use that to bend it.

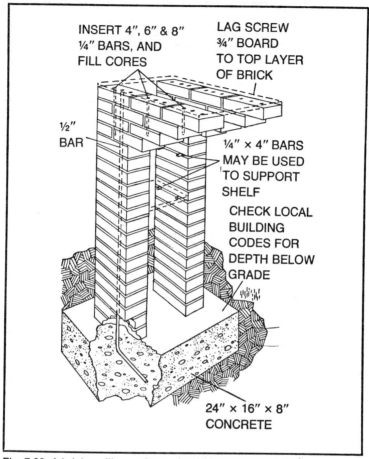

INSERT 4″, 6″ & 8″
¼″ BARS, AND
FILL CORES

LAG SCREW
¾″ BOARD
TO TOP LAYER
OF BRICK

½″
BAR

¼″ × 4″ BARS
MAY BE USED
TO SUPPORT
SHELF

CHECK LOCAL
BUILDING
CODES FOR
DEPTH BELOW
GRADE

24″ × 16″ × 8″
CONCRETE

Fig. 7-30. A brick mailbox column should last nearly forever.

Use leftover bricks or 2 by 4s to set the first leg in place on the poured concrete base. Now set the next post and use a 2 by 4 as a wedge to get the posts the correct distance apart. This will vary depending on the size of your mailbox. Use a 2-foot level to plumb the posts. The posts may actually be set in place, using the bent bar as a support in the excavation along with 2 by 4 bracing and concrete poured around them, if you prefer. The posts must now sit in place for at least two weeks to allow the footing to cure. During this time the tops of the posts should be covered and the footing kept damp (for the first seven days, at least).

The mailbox is installed by driving wooden plugs into the unfilled brick cores. Then lag screw a ¾-inch pressure-treated plank to the wood plugs. The plank should be the same size as the mailbox base. Use pressure-treated 1 by 2 inch furring strips to make a spot to attach the mailbox, which is then set in place over the furring strips. Nail the mailbox flange to the furring strips.

STEPPING STONES

Stepping stones placed 4 to 6 inches apart can provide an attractive looking walkway of practical vlaue. For the stones described here, use solid brick units 3 3/3 inches by 2½ inches by 8 inches. Sixty of these will make 10 stepping stones when combined with 10 lengths of 1 by 4 inch by 6-foot pressure-treated pine, about a pound of 5 penny galvanized nails, 3 cubic feet (about 260 pounds) of sand and a single bag of Portland cement.

For each step, cut a square hole 4 inches deep and large enough to accept the frame's 17½-inch outside dimensions. Set the frame into the hole so that its top is level with the grass or other surface. Now, dry mix 1 part of cement to 3 parts of sand and spread the dry mixture in the hole about 1-½ inches deep. Tamp the mix and lay the brick units in the pattern shown in Fig. 7-31. Make sure the units are flush with the top of the frame. Ten squares such as this should go in place in about two to three hours.

RETAINING WALL

Steep hillside lots tend to erode quickly when not covered with grass. Keeping grass cut on these lots can be a terrible chore.

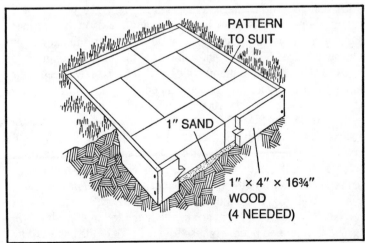

Fig. 7-31. Stepping stones are easily built.

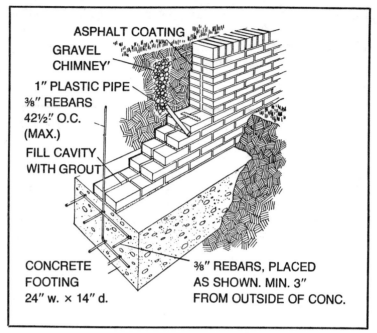

ASPHALT COATING
GRAVEL CHIMNEY'
1" PLASTIC PIPE
⅜" REBARS 42½" O.C. (MAX.)
FILL CAVITY WITH GROUT
CONCRETE FOOTING 24" w. × 14" d.
⅜" REBARS, PLACED AS SHOWN. MIN. 3" FROM OUTSIDE OF CONC.

Fig. 7-32. This sturdy retaining wall will require a lot of work, but it is of straightforward construction.

Retaining walls can be used to terrace such lots or to correct erosion where it has already taken place. Because strength is needed, a double brick wall is necessary. The complex construction of the 8-inch wall covered earlier isn't always needed. The wall we will describe here uses two wythes of running bond, with a header course laid only as a top course. Metal ties hold everything together in conjunction with metal reinforcing bars for added strength.

A check of local building codes will tell you how deep your footing must go. Once that is determined, you can begin to figure material needs. For every 100 square feet of retaining wall, you'll need 1310 3¾ by 2¼ by 8-inch brick. Each foot of wall length requires 4.4 of these brick. For the 100 square feet of wall you'll need 20 cubic feet of mortar, with 2.33 cubic feet of concrete for each foot of wall length. Steel reinforcing bars in ⅜-inch diameter are used in 53-inch lengths and are bent at a 90 degree angle 9 inches from one end. You'll need one of these for every 3½ feet of wall erected. Half-inch steel bars are used for the footing, as shown in Fig. 7-32. You need 5 feet of these for each foot of the base. Allow for a 10-inch overlap. Plastic tubing with a 1-inch outside diameter is used to form drains. You'll also need some asphalt, though not a great deal.

The wall is to be no more than 3 feet high. Metal ties are to be used between wythes, with one placed on each course. The footing is dug out and poured using scrap brick to hold the bottom reinforcing bars off the ground. The footing is 2 feet wide by 14 inches deep, and should be allowed to cure for at least a week before bricklaying is begun. As the wall goes up, cut the bricks where the plastic pipe for the weep holes is to pass through. There should be a weep hole placed every 4 feet along the wall. Before placing the top capping row on edge, pour grout (mortar thinned with water) down around the ⅜-inch reinforcing bars to form a good bond.

Cap the wall with a solid row of brick laid on edge. Do a good job of pointing the joints on both sides of the wall. Now take the asphalt and go along the back edge of the wall, coating it well. If the area is especially damp, install a French drain at the level of the weep hole tubes. A French drain is a bed of gravel several inches deep laid in a trench and then covered over.

MORTARED BARBECUES

Placing a *mortared barbecue* is a bit more of a job than setting the earlier described unmortared unit (Fig. 7-33). For that reason,

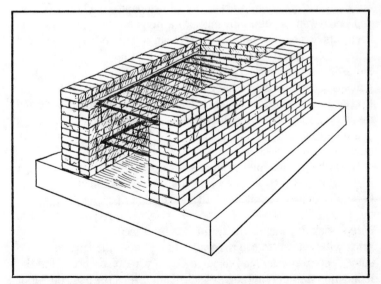

Fig. 7-33. A mortared barbecue along this line is a relatively simple job and provides a permanent outdoor cooking unit.

Fig. 7-34. Serpentine walls are attractive and easily built.

it's a good idea to place it for convenience while also keeping an eye on prevailing winds. You wish to prevent three problems caused by winds. First, you don't want smoke always blowing in the cook's eyes. Second, you don't want smoke blowing into your home. Third, you don't want smoke blowing into a neighbor's home. The barbecue itself need be nothing more than a simple fire pit, with a gravel bottom and walls just high enough to hold a grill. You can build this sort of barbecue in a very short time. Place a 2-inch tamped sand base under the entire unit and pour 4-inch footings. Run up a double thickness (two brick) wall using a running bond and metal ties. Finish the job with a header course of brick laid on edge in a solid row. Line the fire pit with gravel from 1½ to 2½ inches in diameter and start cooking.

All the bending over while cooking on such a fire pit might get tedious, though, so we'll look at a slightly different kind of barbecue. To build this unit, you will need to pour a base 18 inches deep by about 4 feet square. The overall project will require about 400 cored brick 3¾ inches by 2½ inches by 8 inches. About 65 solid brick will be used to cap the walls. In addition, the job requires 6 cubic feet of mortar and about a cubic yard of concrete for the foundation. Reinforcing bar, ⅜-inch diameter, can be used on 1-foot centers to strengthen the foundation. The grills should be set on

⅜-inch reinforcing bar cut to 6-inch lengths and inserted at the proper spots in the joints between the bricks.

After the base has had a week or two to cure, start laying up the corners of the barbecue. First, though, make an outline. Draw on the foundation and do a trial run of dry laid brick on the first course. The bottom course must be well bonded to the base with a good mortar bed. You may wish to lay loose firebrick in the ash pit, though it isn't essential. Once the corners have gone up three or four courses, begin filling in the walls. Including the cap bricks, the total number of courses for this barbecue is 15, providing a height, with ½-inch joints, of about 42 inches to the top of the outside walls. The grills are set in at different heights, with the charcoal holder at the top of the fifth course and the cooking grill three courses higher. You can also set in an extra row of grill holders one course down for food which you may want to cook more quickly. As the sides and back go up, check with your level to see that the tops of the courses are level and the wall is plumb.

OTHER PROJECTS

Let your imagination go. You can build serpentine walls by moving bricks slightly out and narrowing the joints at the front,

Fig. 7-35. Varied patterns can be used for screen walls, but careful planning is required to make sure the units come out right.

while widening them at the back and vice versa (Fig. 7-34). You can erect full brick screen walls with a bit of advance planning (Fig. 7-35). Build your child a sandbox, or create an original barbecue design. A retaining wall can edge a patio and provide the back wall for a barbecue. You can build brick piers to hold a small roof and have a gazebo on your lawn. The only limiting factors in brick outdoor projects are the time and money you wish to spend and your imagination.

Chapter 8
Pole Buildings

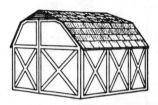

Pole buildings have for years been popular on farms and are now being used for homes in some cases. Part of the reason for their extensive use is the basic simplicity of erection. Instead of digging a foundation, pouring footings, building up foundation walls and so on, you simply dig the approriate number of holes, insert the poles, plumb and brace them and finish the construction.

LOCATING BUILDINGS

No construction of a building of any real size is quite that simple. Pole construction can be used for any type of building, with almost no limits on final size. The larger the building to be erected, the more carefully you must examine the site, select the materials and do overall planning. As buildings increase greatly in size, so do pole length and weight and hole depth and diameter. The difficulty of setting the poles in place is increased. As an example, even poles as short as 15 feet in total length will require a 4-foot deep hole. Poles up to 30 feet need at least a 5½-foot deep hole. At least 6 inches more depth is needed in soft and sandy soils.

Locating pole buildings is similar to locating any building. Because they are often used as barns and loafing sheds, pole buildings are frequently placed so that any open side faces south or east so that they are pretty well protected from prevailing winds. Loafing sheds are simply places for unstabled livestock, most often horses, to retreat from severe weather.

SITE LAYOUT

Poles with top diameters of 5 or 6 inches can be and usually are spaced 12 or 16 feet apart. Lighter poles for smaller

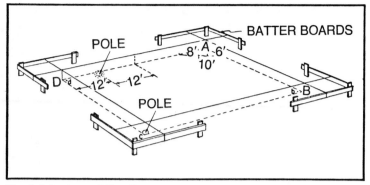

Fig. 8-1. Laying out the site.

buildings should be more closely spaced. In a few cases spacing might be as close as 4 feet, though 8 feet is a more frequent minimum distance.

Lay out the selected site using batter boards, string and stake, just a as you would for any rectangular or square structure (Fig. 8-1). Use the 3, 4, 5 rule to make sure corners are square. In other words, measure 3 feet along one line in a corner and 4 feet along another. If the diagonal is exactly 5 feet, the corner is square. You can use any multiple of these figures you wish, such as 6, 8, 10 or even for large structures, 12, 16 and 20.

SELECTING POLES

Naturally only treated poles should be used. The University of Idaho College of Agriculture states that in ground life of untreated woods is seldom more than 5 years. Pressure-treated wood will provide a life span of more than 30 years. The site should be level or nearly so, with about a 1 percent slope to allow good drainage.

Table 8-1 shows the depths needed for poles of varying lengths. Holes are dug so they are at least 6 to 8 inches larger in diameter than the butts of the poles to be inserted. Poles are placed on a 6-inch bed of gravel. Greatest strength is achieved if you backfill around the pole for at least half its buried height with concrete. Footings in wet, clay-like soils need to be deeper and wider. For a concrete footing in such soil, you may need a diameter as great as 30 inches with a depth of 14 or 15 inches.

Poles selected should be 1 to 2 feet longer than the distance from the base of the pole to the roof. This will allow you to give them a final trim at the plate line.

DIGGING POSTHOLES

It should be obvious that the job of digging post holes for these poles is not one the be assigned to the hand posthole digger. It can be done, especially on small sheds and such, but the work involved quickly becomes incredible. The need to attain depths up to 4½ feet very often means, even for small buildings, that you'll almost have to crawl into the hole with the digger. In most areas of the country, you can rent equipment with which to dig these deep holes. One and two-man power augers can have bore diameters up to a foot. Many will dig well past the depth needed, with suitable extensions. In other areas, you may be able to locate a farmer who has a tractor-mounted posthole digger and is willing to pick up a few extra dollars for digging your postholes. In other areas, you may be able to locate a sub-contractor for the local telephone or power company who has posthole digging equipment of great size and the gear to set the poles quickly and easily. This latter method doesn't exactly come cheaply, but it saves time and effort.

ALIGNING POLES

For the corners of your building, go through the poles carefully and select the four straightest you can locate. These poles are set in place and carefully aligned to establish them in line with ground line, as it is shown in Fig. 8-1. Just enough gravel or dirt is put into the holes at this point to prevent excessive shifting, and the poles are braced. Next the line poles are placed and aligned with the corner poles. Line poles will have the roughest or least even sides placed to the inside of the building. Placing the uneven sides of the poles to the inside of the building cuts down on the distortion of the sidewalls when they are installed.

Table 8-1. Depths of Pole Holes.

TOTAL POLE LENGTH	ORDINARY AND GRAVELLY SOILS
15 feet or less	4 feet
16-20 feet	4½-feet
21-25 feet	5 feet
26-30 feet	5½-feet

For soft soils make the hole ½-foot deeper.

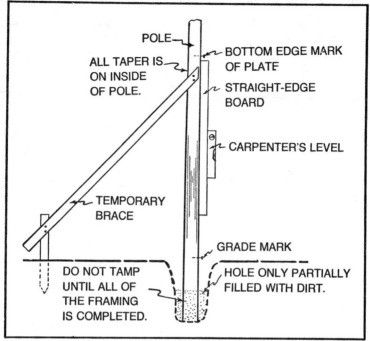

Fig. 8-2. Settings and aligning poles.

Poles are aligned vertically by placing a long straightedge along the outside and using a carpenter's level on that to plumb them (Fig. 8-2). Braces are brought down from near the tops of the poles and nailed to stakes driven into the ground in order to hold the plumb. At this point, get a grade level marking on the poles, keeping it as close to the same level on each pole as is possible. Marks here should be permanent and often consist of a shallow saw cut.

PLACING PLATES

You use the grade marks as distance markers for the plates at the tops of the poles. Accuracy of grade marking will affect the accuracy of your roof line at the eaves, in particular, and at the ridge. If grade marking seems off, you can level the plates as they are installed to correct them. The job goes more rapidly with accurate grade markings. Measure from each grade mark to the predetermined eave line, driving a nail in at this point. The outside plate can now be set on the nails and spiked to the poles. Obviously, the nails must be driven in at a distance equal to the width of the plates used.

202

For greater strength in construction, you can use lag screws, carriage bolts, or notch the poles to receive the plates, so that a tighter unit is formed when nailed. Of course, use only ring or spiral shank nails for the nailing. Figure 8-3 shows the inside plate placed higher than the outside plate so that it, too, will support the rafters. It is installed after the rafters are placed. If your area is one in which heavy snow loads are expected or in which wind loading is likely to be heavy, braces and blocks can be added to help increase the load-carrying ability of the plates (Fig. 8-4).

Purlin plates are placed on interior poles and are used to support the rafters in the same manner as the outside plates and at the same on-center distances. The rafters are lapped and spiked right through into the poles, as you see in Figs. 8-4 and 8-5. Figure 8-5 also shows the notching of the poles used as an alternative method of added support.

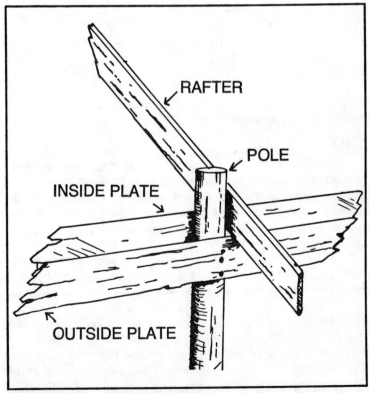

Fig. 8-3. Framing details for the exterior lines of poles.

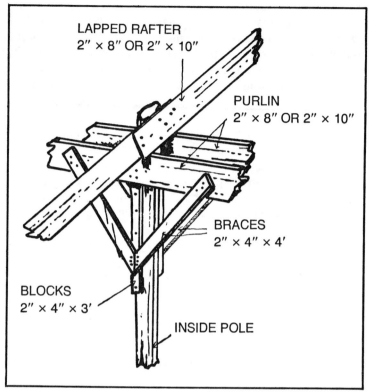

Fig. 8-4. Framing for interior poles.

ERECTING RAFTERS

Once the plates are in place, the rafters can be erected. While truss systems can be used on pole buildings, most often the two sets of end rafters are cut and butted together. The remaining rafters are simply lapped and spiked together at the roof peak so that sawing is almost eliminated. Figure 8-6 shows the method used for interior rafters. When getting material for rafters, remember that you will need one more pair than the total number of on-center distances covered—for the final end rafters. A 48-foot long building with rafters spaced 2 feet on center would require 25 sets of rafters, not 24. Rafters are spaced from 2 to 6 feet on center, depending on the span and the size of the rafter material. Figure 8-6 also shows collar beams in place. Generally, assuming 12-foot pole spacing, you can use 2 by 10 rafters spaced on 4-foot centers. A check of local codes will help, as load factors in your area may differ. For

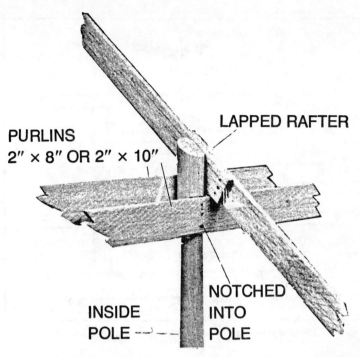

PURLINS
2″ × 8″ OR 2″ × 10″

LAPPED RAFTER

NOTCHED
INTO
POLE

INSIDE
POLE

Fig. 8-5. A second method of framing interior poles, with notches in the poles providing support for the purlins.

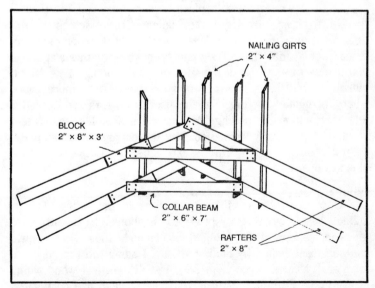

NAILING GIRTS
2″ × 4″

BLOCK
2″ × 8″ × 3′

COLLAR BEAM
2″ × 6″ × 7′

RAFTERS
2″ × 8″

Fig. 8-6. Ridge framing.

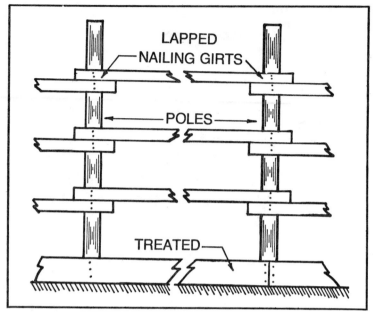

Fig. 8-7. Side wall framing, showing girts lapped to cut down on sawing.

heavy snow loads, I would recommend those 2 by 10s being placed on 2-foot centers, assuming a span of over 20 feet.

Once the rafters are in place, you can saw off the tops of the poles. Now, 2 by 4 girts are nailed in place across the rafters, as shown in Fig. 8-6. These will serve to support corrugated roofing materials. They can be covered with sheathing if you decide to use asphalt shingles or other roofing material requiring a solid base. For rafters spaced 4 feet or more apart, when corrugated roofing is to be used, the girts are spaced 2 feet apart and nailed on edge. For rafters spaced less than 3 feet apart, the girts may be nailed flat, for corrugated roofing. Or 1-inch solid sheathing can be used, if shingles are desired.

SIDING

Once the roof is on, you can move to the siding. This is nailed over 2 by 6 girts spiked to the poles. If the structure gives indications of not being stable enough, these 2 by 6 girts can be strengthened by nailing a 2 by 4 along the top edges to form an L-shaped brace (Fig. 8-7). Usually vertical wood siding is used on pole buildings. But you may wish to use corrugated siding or even some form of plywood sheathing/siding.

Board and batten siding of rough pine is often used, with boards no more than 6 inches wide, especailly if the lumber is green, to cut down on shrinkage and warping effects. Battens are usually of 1 by 2 pine. There are alterations on the board and batten them (Fig. 8-8). Nailing patterns on the siding are important. Boards for board and batten siding are nailed through the center and not at the edges. The batten is nailed through the gaps between boards. This allows the boards to contract and expand without pulling the nails or splitting. For batten and board siding, the battens are single-nailed through their centers and the boards double-nailed. For board and board siding, the under boards are nailed through their centers while the final boards are double-nailed closer to their edges.

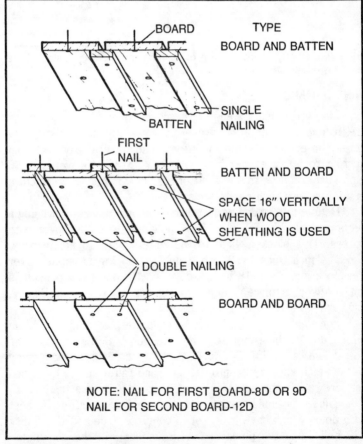

Fig. 8-8. Nailing for vertical board siding.

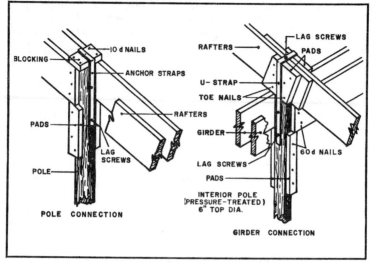

Fig. 8-9. Methods of fastening roof framing members for areas where severe wind can be expected.

WIND DAMAGE

In a great many areas of the country, extreme high winds are frequent happenings. Recently even my area, not really noted for extreme wind conditions, had several days of wind gusting to 50 miles per hour (mph), with a day's worth of 70 mph gusts. It pays to add a bit of bracing to your pole structure in areas where such winds occur often.

Double rafters can be fastened as shown in Fig. 8-9, using a section of blocking on top of the pole and extending over both rafters. This block is held to the rafters with 10 penny nails. Next, ¼-inch thick by 2-inch wide U-strapping is applied over the block and about 18 inches down each side of the pole. It is lag screwed in place, using two lag screws per side. Finally, pads are cut and placed under the rafters.

For single rafter installations, pads are placed and nailed on both sides of the rafter to bring the edges out even with the poles. The same size U-strap is then lag screwed in place, using ⅜ by 3-inch lag screws into the pads and rafter and ½ by 4-inch lag screws into the poles. Pads are then added under the girder.

These methods for preventing wind destruction fit best with poles which have been set in concrete. The added stability will help keep the poles in the ground as will the use of the correct types of nails (Fig. 8-10).

REASONS FOR BUILDING FAILURES

Pole building failure often stems from inadequate fastening and bracing, the settling of poles or the rotting of poles at or below the ground line. The use of gravel or concrete footings beneath poles helps to insure little trouble with settling poles. Utilizing poles at least 8 inches at the butt for load carrying spots also helps. We're assuming here a moderately large building, as sheds seldom need poles larger than 5 inches or so at the butt.

PRESSURE TREATMENT OF POLES

As for poles rotting, whether above or below ground, we've already covered the advantages of pressure treatment of poles and other woods. Little more is needed here to help you extend the life of your pole building to 30 or 40 years. A few features of the various types of pressure treatment might be considered, though. Poles are available treated with all three types of preservative. If you wish to paint your structure after it is up, avoid any of the oil or creosote type preservatives. None of these take paint at all well. Stick to the waterborne preservatives if cleanliness is of importance. Heavy oil type preservatives tend to creep along nails and other fasteners, eventually resulting in the staining of siding or finish lumber in contact with the treated lumber. Rubbing against a cresoted pole is not good for clothing, nor for the skin. The chromated copper arsenates such as those used by Koppers Company to produce Wolmanized and Outdoor woods do not leach out or brush off.

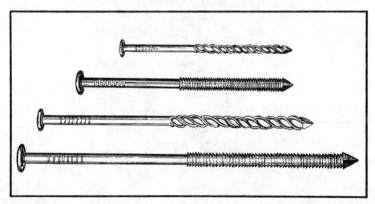

Fig. 8-10. The use of spiral or ring shanked nails is recommended to increase strength.

Fig. 8-11. A 24-foot pole style hay shed can be constructed in units of 12 feet, with a 16-foot eave height.

Fig. 8-12. This pole style bunker silo is good for areas of heavy rainfall and can be easily converted for self-feeding.

Fig. 8-13. For machinery storage, this pole building provides a shop 24 by 30 feet, with a 30-foot wide storage area which can be built in multiples of 12 feet.

Fig. 8-14. A 39-foot pole type storage barn.

Make sure not to cut the butt ends of the poles to be set in the ground. Cutting doesn't destroy the preservative impact, but that end grain has received an exceptionally high loading of the material. End grain is the starting point for most water entry below grade and subsequent rot. Keeping the most heavily preserved end grain intact is only common sense. Figures 8-11 through 8-14 show various pole buildings.

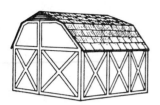

Chapter 9
Silos

To most of us, *silos* are nothing more than those tall slender buildings in which a farmer stores grain. For a farmer, a silo is much more than a conspicuous building put up near his barn. It is a lifeline for his animals through the winter. Back in 1875 there were no silos. Today there are more than 900,000 silos used to store nearly 100 million tons of silage. In many areas, farmers use below ground silos of very simpe structure. In Virginia where I live the most popular method seems to be to simply pile the silage on the ground and thoroughly cover it with heavy duty black plastic held in place with old tires. It doesn't look very nice, but the device certainly seems effective. Such temporary trenches and on-ground silos cause a higher silage loss rate than do upright permanent models. Stack silos may lose 50 percent of the silage, while an upright silo may lose only 5 percent. All horizontal silos appear to have higher silage loss rates.

Still, horizontal silos do offer advantages. It is a simple matter to turn a horizontal silo into a self-feeder simply by using a movable feeding gate. Mechanical feeding is also easier. Tower silos are somewhat more dangerous than horizontal silos because of the buildup of gases as the silage ferments. All fermenting silage produces carbon dioxide, which is a suffocating gas. Some silage produces nitrogen dioxide which is poisonous. Such problems also exist with trench and pit silos, since the nitrogen dioxide, at room temperature, is 2½ times as heavy as air. It is also an orange yellow color, though as the temperature rises its color becomes darker and its density drops. The gas will collect and remain in any depressed area, so care on entering silos is essential.

A starch-iodide paper, readily available from many drug stores and chemical supply houses, can be used to test for nitrogen dioxide. It will turn blue if the air has this gas in much quantity. To test for carbon dioxide, lower a kerosene lantern into the silo. If the lantern goes out, the air must be stirred to dilute the gas. Nitrogen dioxide gas should disappear within a week of filling the silo, but is also present during filling.

TOWER SILOS

The general construction of tower silos includes a need for airtight walls and tight fitting doors. Walls need to be strong, smooth, and circular in form for strength and to permit the silage to settle freely. The walls will also need to be protected, on the exterior, to prevent deterioration. The foundation must be well-drained and strong enough to support a considerable load, or the walls will settle and crack. There should also be a chute to ease feeding and a strong ladder to give access to the filling door. The ladder should also have safety rings. A roof is needed.

The silo should be located as near as possible to the feeding area and should be sized so that no less than 3 inches of silage a day will be removed (to prevent spoiling). This means that overbuilding is always a bad idea with silos. You must determine how much feed your livestock will need and then build accordingly.

Silage Information

If we assume you are using what is considered normal corn silage, a cow feeding on this alone will need about 100 pounds per day during winter. For summer feeding in droughts and for fall feeding, the amounts may be much less—down to as little as 10 pounds per cow a day.

Table 9-1. Relation of Herd Size to Silo Diameter for Winter Feeding.

INSIDE DIAMETER OF SILO (FEET)	VOLUME PER FOOT OF DEPTH	AMOUNT TO BE REMOVED—				ANIMALS THAT MAY BE FED WITH A DAILY ALLOWANCE OF—			
		DAILY	FOR FEEDING PERIOD OF—			80 POUNDS	60 POUNDS	40 POUNDS	20 POUNDS
			120 DAYS	180 DAYS	240 DAYS				
	Cubic feet	Pounds	Tons	Tons	Tons	Number	Number	Number	Number
8 ------	50.3	419	25	38	50	5	7	10	21
10------	78.5	654	39	59	78	8	11	16	33
12------	113.1	942	57	85	113	12	16	24	47
14------	153.9	1,282	77	115	154	16	21	32	64
16------	201.0	1,675	101	151	201	21	28	42	84
18------	254.5	2,121	127	191	255	27	35	53	106
20------	314.2	2,618	157	236	314	33	44	65	131

Basis: 50 pounds of silage per cubic foot; 2 inches removed daily to avoid spoilage.

INSIDE DIAMETER OF SILO (FEET)	VOLUME PER FOOT OF DEPTH	AMOUNT TO BE REMOVED—				ANIMALS THAT MAY BE FED WITH A DAILY ALLOWANCE OF—			
		DAILY	FOR A FEEDING PERIOD OF—						
			60 DAYS	90 DAYS	120 DAYS	80 POUNDS	45 POUNDS	20 POUNDS	10 POUNDS
		Pounds	Tons	Tons	Tons	Number	Number	Number	Number
8	Cubic feet 50.3	629	19	28	38	8	14	31	63
10	78.5	981	29	44	59	12	22	49	98
12	113.1	1.414	42	64	85	18	31	71	141
14	153.9	1.924	58	87	115	24	43	96	192
16	201.0	2.513	75	113	151	31	56	126	251
18	254.5	3.181	95	143	191	40	71	159	318
20	314.2	3.928	118	177	236	49	87	196	393

Basis: 50 pounds of silage per cubic foot; 3 inches removed daily to avoid spoilage.

Assume a herd of 20 cows. Your plans might include feeding 60 pounds of silage per day for as long as 240 days. That's 144 tons of silage for the year. Allow 10 percent for loses, and your silo will need to be 14 feet in diameter and 45 feet high. Table 9-1 gives the herd size/silo diameter needed for winter feeding, while Table 9-2 provides the same figures for summer feeding. Table 9-3 shows the capacities of silos with different diameters and heights. All these are rough estimates, as silage will vary in weight. Feeding conditions will also change, depending on the type of silage and the area of the country. Se-

Table 9-3. Capacity of Silos With Different Diameters and Depths of Silage.

DEPTH OF SILAGE (FEET)	CAPACITY WITH AN INSIDE DIAMETER OF—						
	8 FEET	10 FEET	12 FEET	14 FEET	16 FEET	18 FEET	20 FEET
	Tons	Tons	Tons	Tons	Tons	Tons	Tons
20	18	27	-----	-----	-----	-----	-----
22	20	30	-----	-----	-----	-----	-----
24	22	34	49	-----	-----	-----	-----
26	25	38	55	-----	-----	-----	-----
28	28	43	61	84	-----	-----	-----
30	31	47	68	92	121	-----	-----
32	34	51	74	100	131	-----	-----
34	-----	56	80	109	142	180	-----
36	-----	-----	86	117	153	194	-----
38	-----	-----	93	126	165	209	258
40	-----	-----	100	135	177	224	276
42	-----	-----	-----	144	188	237	293
44	-----	-----	-----	152	198	251	310
46	-----	-----	-----	-----	209	265	327
48	-----	-----	-----	-----	220	279	344
50	-----	-----	-----	-----	231	293	361

Basis: Silo filled with normal corn silage at the average speed of 20 to 50 tons per day; refilled once after silage has settled and the top has been well tramped. For grass silage, capacity is increased from 5 to 10 percent.

vere winters require more silage. Table 9-4 gives estimates of silage weight per cubic foot. Grass silages such as alfalfa, orchard grass and cereal crops weigh about 10 percent more.

Foundation Layout

The silo foundation must go below frost depth for your area, which may mean as deep as 4 feet or a bit more in Northern areas. In no case should a silo of any size have a foundation depth of less than 2 feet. The silo walls will carry as much as 60 percent of the silage weight much of the time.

Table 9-5 provides information on footing width for silos of different height and construction. Concrete block can be used for the below grade foundation walls. Use 12-inch wide block. The footing must rest on undisturbed ground. The table estimates are for normal, firm and dry soil with a load bearing capacity of no less than 2½ tons per square foot. Soils in some areas will

Table 9-4. Weight of Settled Silage per Cubic Foot When the Silo Is Full.

DEPTH OF SETTLED SILAGE (FEET)	AVERAGE WEIGHT PER CUBIC FOOT (POUNDS)	DEPTH OF SETTLED SILAGE (FEET)	AVERAGE WEIGHT PER CUBIC FOOT (POUNDS)
1	17.7	26	46.4
2	23.5	27	46.7
3	26.9	28	46.9
4	29.5	29	47.2
5	31.6	30	47.4
6	33.3	31	47.7
7	34.7	32	47.9
8	36.0	33	48.1
9	37.1	34	48.3
10	38.1	35	48.5
11	39.0	36	48.7
12	39.8	37	48.9
13	40.6	38	49.1
14	41.2	39	49.3
15	41.8	40	49.5
16	42.4	41	49.7
17	43.0	42	49.9
18	43.5	43	50.0
19	43.9	44	50.2
20	44.3	45	50.3
21	44.7	46	50.5
22	45.1	47	50.6
23	45.5	48	50.8
24	45.8	49	50.9
25	46.1	50	51.0

Basis: Corn silage (moisture content, 72 percent), as fed out of silo. Grass silage under the same conditions averages from 5 to 10 percent heavier than corn silage.

Table 9-5. Footing Widths for Silos of Different Heights.

HEIGHT OF SILO WALL (FEET)	LOAD AND WIDTH OF FOOTING FOR SILO WALLS OF—					
	6-INCH MONO- LITHIC CONCRETE		2½-INCH CONCRETE STAVE OR 5-INCH TILE		2-INCH WOOD OR METAL	
	LOAD PER FOOT OF CIRCUM- FERENCE	WIDTH OF FOOTING	LOAD PER FOOT OF CIRCUM- FERENCE	WIDTH OF FOOTING	LOAD PER FOOT OF CIRCUM- FERENCE	WIDTH OF FOOTING
	Tons	Inches	Tons	Inches	Tons	Inches
20----------	1.5	8	1.0	8	0.8	8
30----------	2.5	12	2.0	10	1.5	8
40----------	4.0	20	3.0	15	2.7	13
50----------	5.7	28	4.4	22	------	------
60----------	7.8	38	6.5	32	------	------

require larger footings, and this can be determined by checking with your local building department.

Foundation layout is simplified by the use of a 2 by 4 cut a little longer than half the diameter of the silo, plus the width of the foundation wall. Drive a stake into the ground at the center point of the silo. Then drive a 40 penny spike through one end of the 2 by 4 and into the stake. Now measure off the distance on the 2 by 4 to the outside of the footing and nail on a marker. See Figs. 9-1 through 9-4. For building sites that are not so level, use a slightly longer marker and hold the 2 by 4 level as you make your circle. For extremely uneven ground, you can touch down as often as possible (every few inches is best) and then join the marks later. Now move the marker in on the 2 by 4 the amount needed to draw the inside line for the footing. If the footing trench is dug cleanly, you should be able to pour directly into it, without using forms. If the soil is so loose a clean hole isn't possible, you will have to provide forms, as shown in

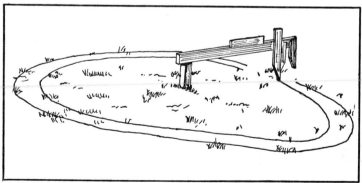

Fig. 9-1. Laying out the foundation for a silo.

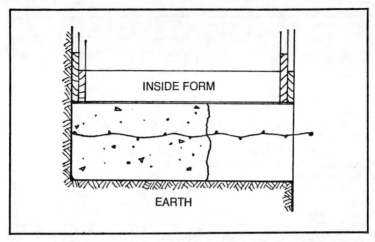

Fig. 9-2. Method of constructing a foundation without footing for a monolith silo in firm soil.

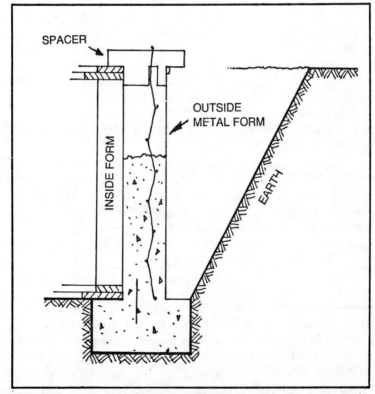

Fig. 9-3. The method used for a monolith silo when a footing is required.

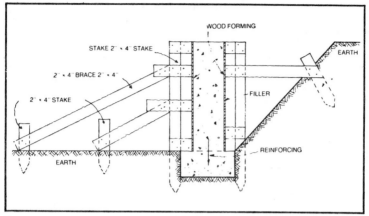

Fig. 9-4. Built-up forms for foundation walls.

Fig. 9-5. Though this shows a foundation form above ground, the footing form is made in the same manner below grade.

Excavation will have to be large enough to allow installation and removal of the forms. Transit-mixed concrete is probably the easiest way to go, but you may wish to machine mix your own. A mix of 1 part Portland cement, 2½ parts of sand and 4 parts of gravel will be adequate for small to moderate size silo foundations and footings. The gravel should be clean and range in size from ⅜ inch to 1½ inches. Table 9-6 gives proportions for mixing approximately a cubic yard of concrete at the site.

Floors and Walls

Silo floors can consist of the bare ground or concrete. Concrete floors will add considerably to the cost of the silo, but will also keep rats from burrowing in from beneath the footings. Figure 9-6 shows the drain layout needed for various types of floors, allowing silage jucies to drain properly.

All silo walls need to be reinforced, but the height of the silo and the weight of the silage determine how much. Frequently 9/16-inch hoops are used, drawn up just tight enough for several turns of the threads to show beyond the nut. Table 9-7 gives the hoop spacing for various distances down the sides of the silo. Note that the closer to the top you get, the greater is the spacing allowed. This spacing is designed to allow you to use silage with a water or moisture content up to 74 percent. For longest life, glavanized hoops are a good idea.

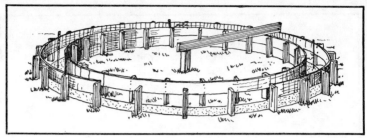

Fig. 9-5. Forms for pouring the footing and foundation extend partly above ground.

Doors and Doorways

Silo doors must fit tightly to reduce air flow as much as possible, but there is also another requirement. They must be flush on the inside or the silage will settle unevenly, creating air pockets. Door frames must be very strong to resist pressure. Most doorways for a silo will be of the intermittent variety, as continuous doorways are very difficult to frame strongly enough. For intermittent doorways a good size is 20 by 30 inches, with the doorways spaced about 3 feet apart.

CONCRETE STAVE SILOS

If you decide to build a concrete silo and wish to do the work yourself, avoidance of the *monolithic*, or one piece, concrete silo is probably a good idea. Form costs and labor costs for assistants can be murderous. Concrete staves are readily available and, for moderate size silos, can be erected with much less help. The staves are held in place with the metal hoops already described. Chutes for these silos are usually of concrete

Table 9-6. Proportion and Approximate Quantity of Materials Required for Making 1' Cubic Yard of Concrete In Place.

PROPORTION			QUANTITY		
CEMENT	SAND	GRAVEL OR STONE	CEMENT	SAND (DAMP AND LOOSE)	GRAVEL (LOOSE)
			Sacks	Cubic yards	Cubic yard
1	1.5	--------	15.5	0.86	--------
1	2.0	--------	12.8	.95	--------
1	3.0	--------	9.6	1.07	--------
1	2.0	3.5	6.5	.48	0.84
1	2.5	4	5.6	.52	.83

Quantities may vary as much as 10 percent, depending on the aggregate.

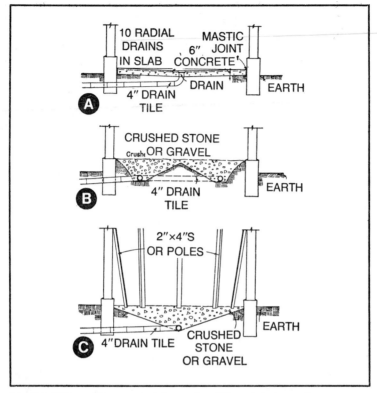

Fig. 9-6. Different floor materials require different drainage forms.

staves and are bound to the main structure with supplementary hoops. When the staves are in place and braced and the hoops placed and tightened, concrete stave silos are ready to have the joints sealed. Then the interior surface is finished with a cement plaster and painted with a plastic sealer. To aid in erecting concrete stave silos, you can also buy precast concrete door frames.

METAL SILOS

Metal silos go up easily and can give good service if cared for properly. Usually the required care is an annual painting of the interior with raw linseed oil to prevent corrosion from the acids in the silage. Metal silos need to be anchored with guy wires to prevent wind damage when they are empty. It is usually a simple matter to locate these silos (they come in complete packages) in rural areas. From that point it's simply a matter of carefully following the manufacturer's erection directions.

WOOD STAVE SILOS

Wood stave silos also come as packages in most cases and are relatively easy to erect. Wood offers an advantage for storing silage, since it is unaffected by the silage acids. But wood suffers from a tendency to decay unless pressure-treated and will swell and shrink as it takes on or loses moisture. This can lead to damage if the hoops are not loosened during loading. Loose hoops can allow wind damage when the silo is empty. Pressure treatment will just about cut out other maintenance needs.

WOODEN HOOP SILOS

Probably the most sensible for small farm use and overall ease of construction is the wooden hoop silo. The silo is made of hoops of thin boards lined with plywood, or even 1-inch tongue and groove pine sheathing, in a double thickness with 15 pound roofing paper between them.

To start a wooden hoop wall, foundation anchors are embedded in the concrete or concrete block wall. Depending on the diameter of the silo, you will need four to six anchors of ⅜ by 1½-inch strap iron, about 46 inches long. Each piece will have the lower end bent out 2 inches, which is the portion to be embedded. Two holes are drilled in the upright portions for ½-inch bolts, with one hole 2 inches and one 2 feet from the upper end of the strap. Anchor irons are placed in the foundation wall 2¾

Table 9-7. Spacing of 9/16-Inch Hoops for Stave Silos.

DISTANCE OF HOOP BELOW TOP (FEET)	SPACING FOR A SILO WITH A DIAMETER OF—					
	10 FEET	12 FEET	14 FEET	16 FEET	18 FEET	20 FEET
	Inches	Inches	Inches	Inches	Inches	Inches
0 -20	30	30	30	30	30	30
20-22½	30	30	30	15	15	15
22½-25	30	30	30	15	15	10
25-27½	30	30	30	15	15	10
27½-30	30	30	15	15	10	10
30-32½	30	30	15	15	10	10
32½-35	30	30	15	10	10	7½·
35-37½	30	15	15	10	7½·	7½·
37½-40	30	15	15	10	7½·	6
40-42½	------	15	15	10	7½·	6
42½-45	------	15	15	10	6	5

All hoops to have rolled threads and the next larger size of nuts. Use ⅝-inch hoops if 9 16-inch are not available.
Figures in this table are based on a safe working stress of 20,000 pounds per square inch (for silages containing up to 74 percent of moisture).   For silages with a moisture content higher than 74 percent. provide vertical drainage channels on the inside surface of the silo walls.

inches farther from the center of the silo than the inside of the silo wall, with the lower hole 4 inches above the foundation. The anchor irons are bolted through the two lower hoops to the silo wall.

Hoop Form

Hoops are built on a form, which is started by driving a 2 by 4 stake into the ground and sawing it off a foot from the ground. Fasten one end of a strip a few inches longer than the radius of the silo to the top of the stake, using a 10 penny nail. Measure from the nail out along the strip to a point 1 inch longer than the radius of the silo. Cut the strip off at that point. One inch toward the center from the outer end of the strip, drive a 1 by 4-inch stake about 20 inches long. Swing your strip halfway around the circle and drive a similar stake in line with the first stake and the center stake. Swing the strip a quarter of the way around and drive another stake. Keep driving stakes around the circle until the distance between stakes is no more than 2 feet. To these stakes, you now fasten 2 by 4 uprights, 6 feet long, with the outer edges plumb and flush with the end of the measuring strip. Once all the uprights are neatly placed, you can remove the measuring strip.

Opposing uprights are tied, at the bottom, using 1 by 1-inch strips of material. The strips are cut to the exact diameter of the form which will be the same size as the outside diameter of the silo wall. Mark these strips in their centers. Nail the lowest one at this point to the top of the center strip, and nail the remaining strips to each other at the middle marks. Beginning about 4 feet from the ground, tie the upper parts of the uprights in the same manner. Plumb the uprights and brace them to each other. Then run four braces from the uprights to the center stake.

Making Hoops

Once this form is finished, you can begin making the hoops. Hoops will be of ½-inch thick wood, preferably 4-inch wide pressure-treated pine. Most hoops for moderate size silos will be 3-ply. Make a mark on one of the uprights 6 inches above the ground. Level around this mark using a line level or a carpenter's level and mark each upright. This will be the upper edge of the first hoop. Fasten one end of the hoop wood to one of the uprights so it can be bent around the form to the left.

Nail it to each upright, keeping the upper edge straight flush on the marks. At the third upright, you begin the second ply of the hoop, and at the sixth you start the third ply. Bend the boards on around the form, nailing to each upright with 8 penny nails. Use 6 penny nails between uprights and at joints. Fit butt joints in the outer layer carefully. The second hoop is started in the outer layer carefully. The second hoop is started on the stud to the left of the point where the first hoop began.

You'll need half the number of hoops as the height of the silo plus one. A 30-foot silo will require 16 hoops, while a 20-foot one will need 11. Once the hoops are completed, they're numbered from the bottom up. If you draw four perpendicular lines on each hoop, it will help you later to get things plumb. The form is now torn out and the nails clinched. Joints should be protected from weather with pieces of galvanized steel a foot

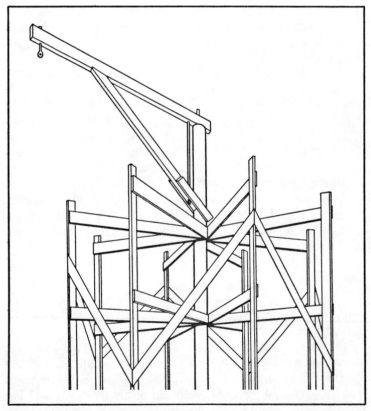

Fig. 9-7. The basic framing of an inside scaffold.

long placed over each one. The hoops are placed on the foundation in the same order and position as they were on the form. Make sure one of the perpendicular lines is immediately to the left of the proposed door opening.

Scaffold

The inside scaffold has to go up to provide a safe work area (Fig. 9-7). While still on the ground, splice four staves, each equal to the proposed height of the scaffold, and mark them for hoop spacing. Make the marks for the first hoop 6 inches from the staves' bottoms. The next three marks are 23 inches apart and the rest are 24 inches, except that the last two hoops will be 23 inches apart. The staves are now placed in position on the foundation, each flush with one of the perpendicular lines on a hoop. The staves are temporarily nailed to the bottom hoops. The top hoop is raised with ropes and nailed to the tops of the staves, using 2 by 4 bracing on the scaffold to keep things in place. Now the hoops are plumbed at these spacing staves and braced to the scaffold.

The first stave is carefully placed and plumbed to the left of the door opening. The staves are nailed as flooring is, driving the tongue into the groove. Each stave is blind-nailed to each hoop with two nails (Fig. 9-8). Joints must be staggered. It is a good idea to check the plumb every few feet around the wall. Continue on around the silo until you reach the opposite side of the door opening, which should be about 35 inches wide in order to accept framing and a door jamb for a 24-inch wide door. The bracing staves are removed as the sheathing progresses to them. As Fig. 9-8 shows, the silo is now bolted to the foundation, using ½ inch by 3⅓-inch bolts that run through staves, hoops and anchor straps.

Doorjamb

The doorjamb is built of 1 by 4-inch lumber 35 inches long, with the ends beveled to fit the inside curve of the hoops. Nail the ends of one or these pieces across the opening on the inside of the upper hoop, and one on the inside of the upper hoop. The remaining pieces are fitted in the same way, placed, carefully plumbed and then nailed. After these crosspieces are in place, nail an additional stave on each side of the door opening. Two staves are prepared for the doorjambs. Remove the tongue from one and the groove from the other. Fasten them in place on

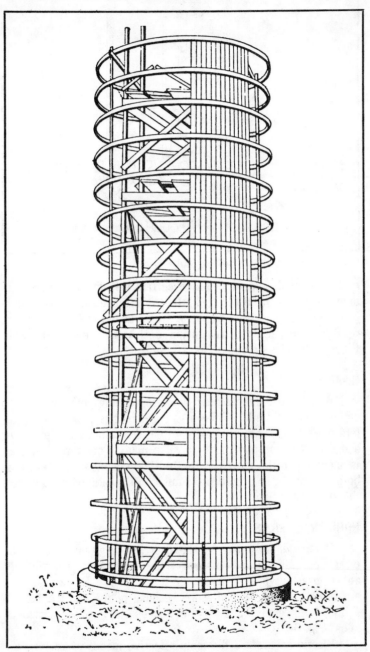

Fig. 9-8. The start of the wooden hoops scaffold, showing the hoops in place, along with the interior scaffolding.

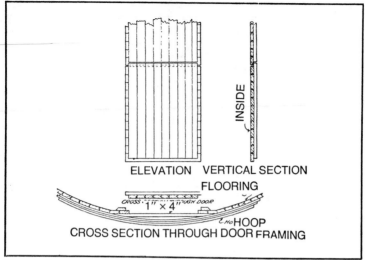

ELEVATION   VERTICAL SECTION

FLOORING

CROSS 1" X 4" PUSH DOOR

HOOP

CROSS SECTION THROUGH DOOR FRAMING

INSIDE

Fig. 9-9. Details of the door and doorway.

each side of the opening. Make certain the finished opening is uniform in size from top to bottom. On the inside of the doorjambs and an inch back from the edges of the opening, fasten 1 by 3-inch strips. Bevel the strips' back edges toward the wall. Figure 9-9 gives details of door framing.

Roofing

A relatively simple pitch roof is probably your best bet to cap a homemade silo. While a gambrel roof is good since it offers more head room, it is also a great deal more complex to cut and nail. You can buy metal roofs in many diameters. The roof is anchored to the staves at the top hoop using straps bolted to both the rafters and the hoop, similar to the method used to bolt the silo to its foundation.

HORIZONTAL SILOS

Horizontal silos may be above or below ground and can be either permanent or temporary. In most cases, the simpler above ground types are considered temporary. Horizontal silos are being used more and more because of the self-feeding advantages. Specially built gates or fences are ideal for horizontal silos (Fig. 9-10).

Plastic covers are the most effective means of keeping the weather from horizontal silos, but they must be well sealed

228

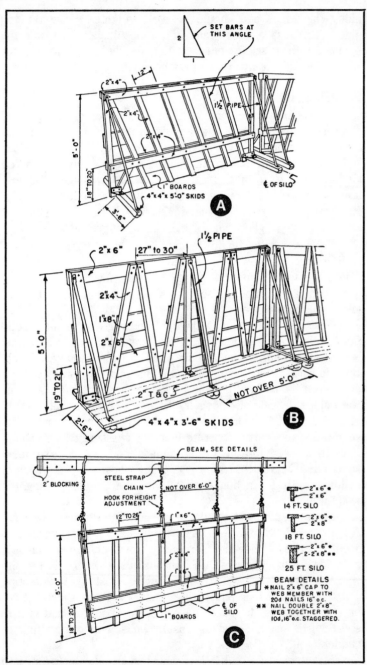

Fig. 9-10. Fence styles for turning horizontals silos into self feeders.

along the edges and should be well weighted along the top sides. The less air that can get to the silage, the lower the spoilage rate. Black 10 mil polyethylene sheeting held in place with many old tires will do a very good job.

BUNKER SILOS

Bunker silos have some advantages over trench, upright and on-ground silos, as well as a few disadvantages. Drainage is not as much of a problem with bunker silos. They're much cheaper and easier to build than vertical silos, and can easily be adapted for self-feeding or mechanical feeding. Top spoilage is almost always a bit greater than in vertical silos. Care is needed in packing and distribution of silage to cut down on spoilage. Cost of bunker silos is usually higher than for trench and on-ground silos.

A bunker silo should be built on a reasonably level and well-drained site, as close to the barn or feedlot as possible. The site should also have plenty of space for trucks and tractors. Drainage is needed so that water will run off and silage juices will be removed.

A bunker silo is built with tight and rigid walls and a solid floor. Materials can vary with local availability. I've seen a bunker silo of nothing more than piled earth and concrete block. But the walls must be sturdily built to support the weight of not only the material held but the packing equipment used. An outward slope to the walls is desirable. If the silage is well distributed and packed, a slope of 1½ inches per foot is plenty. For your walls, use earth, concrete block or pressure-treated wood. If earth is used for the walls, compacting must be extremely well done. Figures 9-11A through 9-11E show some methods of building walls for a bunker silo, with the simplest being the wood wall.

Posts used must be extra heavy or braced to take the load. A minimum depth of 3 feet for post butts and a butt diameter of about 8 inches is needed. For safety, a wooden guard rail is a good idea. Run the posts 3 to 6 inches above the top of the silo wall. Then run 4 by 4 guard rails along the posts. When erecting a wooden wall, 2-inch plank will be needed. Posts should be set on 6-foot centers. If square edge instead of tongue and groove sheathing is used, cover the inside walls with roll roofing or builder's felt.

Make sure that the floor of your bunker silo won't turn into a mudhole. A concrete floor is expensive, but insures fewer

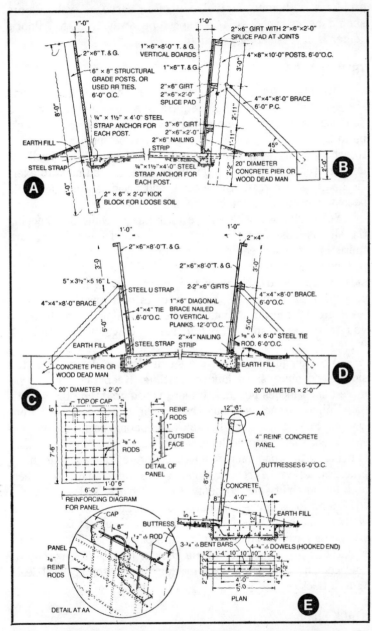

Fig. 9-11. Wall types for bunker silos. (A). Cantilevered post and plank. (B). Braced post and wood girts. (C). Braced steel angle girt with vertical plank. (D). Braced wood girts with vertical planks. (E). Concrete buttress with reinforced concrete wall.

problems with mud. Gravel will often serve as well. Floors should be crowned so that the slope away from the floor's center is at least ¼ inch per foot.

TRENCH SILOS

An unlined *trench silo* is just about as cheap as an on-ground silo, except for the labor required. A few minutes with a bulldozer will produce one of these silos. Spoilage may be high, up in the 25 percent range. Topography is of importance since a trench that fills with water is no help in feeding. Side walls of unlined trench silos may also tend to cave in. You need to exercise particular care during loading and packing. And gas could become a problem in the bottom of the trench.

Select the site with an eye to drainage, while still keeping the silo as close as possible to the barn or feedlot. Make sure there is room for equipment to turn easily when loading or picking up feed. If possible, the trench should be cut into a hillside. Keep the trench away from areas where it might catch seepage from manure deposits, feedlots or barns. A trench on level ground can be cut and should have sloping ends to allow you to drive in and out while filling (Figs. 9-12A through 9-12C).

When digging the trench, a slight slope is desirable on the sides. Too much slope hinders settling. Keep the slope to no more than 4 inches per foot. If more is required, line the walls.

Unlined trenches should have their edges protected from cave-in. Fences will keep livestock and people back.

Trench silos can be lined with a wide variety of materials, starting with wood and going to concrete block as the most practical. Pressure-treated wood is probably today's best bet, both for durability and economy. Cement block is also used. Native stone for a rubble masonry wall can also be utilized. Such walls should be at least a foot thick, using a cement mortar consisting of 1 part Portland cement to 6 parts sand. Then the wall is pargeted, or plastered, with a cement plaster of 1 part Portland cement to 3 parts sand, making the plaster layer at least ½-inch thick.

STACK SILOS

Stack silos are generally considered to be for emergency use. Well-drained and firmly packed land using a gravel floor, or even a cement floor, would be preferred. If the land is well

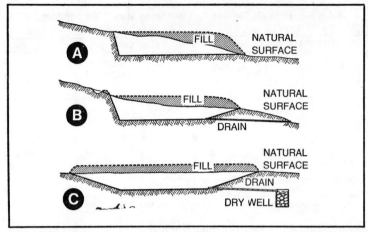

Fig. 9-12. Using the terrain helps when building a trench silo. (A). Hillside. (B). Gentle slope. (C). Level ground.

packed and drained, simply placing the silage on the ground and then covering adequately with black plastic film at least 10 mil thick will do a good job. Losses are about equal to a trench or bunker silo if the cover is well weighted and kept tight.

FENCE SILOS

The use of rings of various kinds of fence to hold silage is common in some areas of the country. For the silo itself, you can use 2 by 4-inch 11 to 12½ gauge welded mesh (galvanized), or ½ by 1½-inch slat fencing using at least five strands of 2-wire cable to join the slats. The diameter, as shown in Table 9-8, can be as much as 24 feet. The total height of the fence silo should never go beyond two 4-foot sections of fence, overlapped at least a foot. Unlined fence silos are sometimes used. But spoilage is very high. Lined fence silos are more practical, as a 4-6 mil black polyethylene sheet liner and cover can reduce spoilage by at least 50 percent and also keeps cut silage from dropping through the fence.

Some people go higher with fence silos, but the prospect of losing the top of the load increases. The fence is not really designed to support as much weight as will be applied when the 12 foot or 16-foot high fence silo is filled.

Use the technique for laying out a tower silo, with the center board and 2 by 4 pivot board and marker, to get the correct and true circle. Then trench around the line, making the

Table 9-8. Materials Required for Lined Fence Silos of Different Sizes.

SIZE OF SILO		CAPACITY	FENCING		LENGTH OF PAPER OR PLASTIC FILM
DIAMETER (FEET)	HEIGHT (IN 4 FOOT SECTIONS OF FENCE)		PIECES	LENGTH OF EACH PIECE	
	Feet	Tons	Number	Feet	Feet
8 ------------	3	8-11	3	25	80
10½ --------	3	14-17	3	33⅓	105
12----------	3	19-25	3	37½	120
12----------	4	22-29	4	37½	160
16----------	4	39-52	4	50	212
19----------	5	70-95	5	60	320
24----------	7	160-230	7	75	553

Paper 4 feet wide; quantities shown are for corn silage and should be doubled for grass silage.

trench at least 15 inches wide. Keep the silo on ground as level and well-drained as possible. Fence ends should be lapped at least two spaces and wired as tightly as possible. Make sure no wire ends stick into the silo where they might cut the plastic lining. The first section is filled to within 6 inches of its top before the second section is wired in place. Twine ties will hold it in place until filling starts, at which time wire ties are needed and the twine cut away. Now cover and your silo is ready.

Chapter 10
Small
Frame Structures

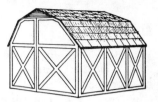

Small frame structures, from sheds to small buildings resembling houses or barns, are still extremely popular. They can, with your labor, be erected with a fair degree of economy. Here we'll cover what are really the fundamentals of small frame building from scratch. Make a careful check of local building codes at the time of applying for a permit to erect the structure. It is sometimes the case that nonresidential structures do not need to meet the same requirements for structual rigidity as do residences. As always, make sure you have permits and know just what is allowed in your locale.

PLANS

Especially in suburban areas a plot plan, showing the location of the existing home as well as the planned outbuilding, will be needed. You will need some clear drawings of the building to be erected. In most cases, the building inspector will accept neatly done scale drawings on graph paper for non-residential structures. If not, you can make your plans and then take them to a draftsman to have professional drawings made. The draftsman can then arrange to have blueprints made, if needed. Be as specific as possible when drawing up your materials list, so the building will meet local codes. His suggestions could prove invaluable at this point, saving a great deal of frustration at a later date.

Also detail on-center spacing, sill size, number of anchor bolts to the foundation and their locations and sizes, sheathing type and thickness, type of roofing materials, type of heat (if any), number of doors, planned use of the building, number of windows and their sizes. Many codes require that at least one window in each room of a residence be a minimum of 24 by 30 inches. If the doorway is blocked by a fire, exit through a window of this size is still possible. Codes probably won't make such a window requirement for out buildings, but it is nonetheless a good idea.

SITING

To gain the most heat from the sun, outbuildings should be sited with most windows and doors facing south or east whenever possible. Try to select a reasonably level site. This makes construction of the pier style foundation easier and also simplifies setting in a concrete slab, if that's your choice. Try to make sure any surface drainage will be away from the building if it is on a slab or has a basement. Make sure property line clearances are correct, and then decide which corner will go where. The next step is to locate the corners.

Start with the first corner and lay out a rough rectangle (or square or L-shape, as your plan dictates). Next, place batter boards, shown in Fig. 10-1, about 3 to 4 feet in back of the wall line. Cut saw kerfs and place three more sets of batter boards at the other corners. At the first corner, drive a stake. Draw mason's line tight and level from saw kerf to saw kerf (Fig. 10-1). Square your first corner by measuring 3 feet up one line (from the corner, not the batter boards), 4 feet up another and checking the diagonal. If the diagonal is 5 feet, then the corner is square. If it isn't, you'll have to move the batter boards until the corner squares up. Repeat the process at all corners. For larger structures, you can go ahead and use larger multiples.

For basements, the excavation can now be made down to footing depth. Use the highest elevation of the excavation's outline to establish the depth of the foundation, as this should aid in providing good drainage. You can use foundation walls no more than 7 feet 4 inches high. For greater head room, 8 feet is better. If heat and plumbing is to be installed, I would recommend another course of concrete block to bring the total to 8 feet 8 inches. Foundation walls are extended above grade to provide good protection from termites and rot. I would also recommend using both a sheet metal termite shield and pressure-treated lumber when laying the sills.

Digging a hole for a basement is something hardly anyone does by hand anymore. A bulldozer or backhoe is quicker and easier on the back and arms. The footing excavation should be made just before the footings are to be poured, so that weather won't cause the sides to crumble and collapse and force you to

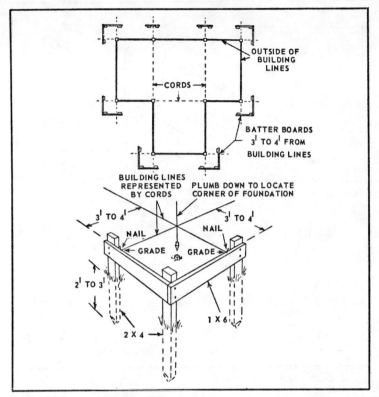

Fig. 10-1. Batter boards.

build footing forms. You may have to build forms anyway if the soil is sandy or soft.

FOOTINGS

Footings for a concrete block basement wall are twice as wide as the wall is thick. Eight-inch block requires a 16-inch wide footings. The foorings are set below frost level always. No footing should be less than 6 inches thick.

Pier and post footings should be square with a post or pedestal (unless you are building up square piers of concrete block). Commonly, pier footings will be 24 inches by 24 inches by a foot thick. Softer soils or wider spacing might force you to go to 30 inches by 30 inches by a foot thick or even larger. Pier footings can be poured or bought precast and set.

Spacing of piers depends on several things, starting with the overall size of the building. Obviously, each corner will re-

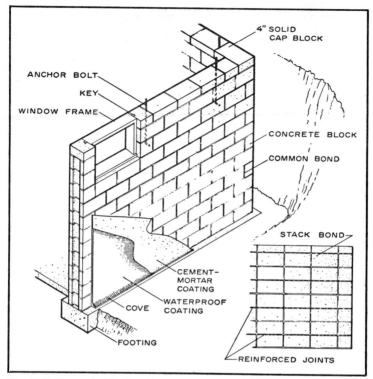

Fig. 10-2. Concrete block walls.

quire a footing. For a small shed, this may be all you need. Footings for larger structures should be spaced as needed, usually on 8-foot centers, with lumber of the correct size for the overall span.

Poured pier foundations are fine for areas with rock shelves partly under the proposed building. The rock is cleaned well and a special bonding agent is used to make sure the pier cannot shift.

CONCRETE BLOCK WALLS

While concrete block comes in a wide variety of styles and sizes, the most common sizes are 8, 10 and 12 inches wide. The 8-inch block is probably about the only one of interest to us here, since loads on the walls for outbuildings will seldom be great enough to require thicker walls. These blocks are nominally 8 inches by 16 inches by 8 inches, with actual measures of 7⅝ inches high and 15⅝ inches long to allow for mortar joints.

Running bond is often used—in fact, I can't remember ever seeing a concrete block wall done in another form. Joints are tooled to prevent water penetration.

Cap blocks are often used on basement walls. The anchor bolts for the sill will be placed through the top two rows of block and through the cap, using a large plate washer as an anchor at the bottom of the block. Block openings around anchor bolts are filled solidly with mortar. For walls to be backfilled, use an asphalt compound to further cut down on seepage. Where especially wet conditions prevail, use either polyethylene sheeting (2 to 4 mil) or roofing felt set into a light asphalt coating with at least 6-inch laps. Then coat the wall with hot tar or asphalt (Fig. 10-2).

ANCHOR BOLTS

Sill plates must be anchored to the foundation wall using ½-inch bolts. Do not use masonry nails as these simply do not have the holding power needed in most areas. For general use, space anchor bolts 8 feet apart all around the foundation wall, unless your local codes specify otherwise.

After the wall is up and the anchor bolts are in, place a sill sealer of fiber glass (you can use unbacked 3½-inch fiber glass insulation) along the wall to make up for any irregularities in the concrete. On top of the sill sealer, especially in areas of heavy termite activity, I would recommend that you place aluminum

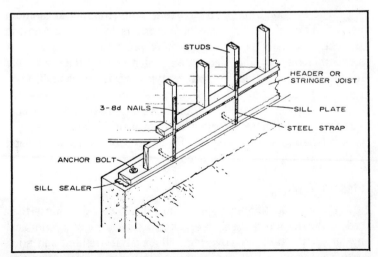

Fig. 10-3. Anchor bolts in place.

flashing cut 2 inches wider than the wall. Keep it at least ½-inch away from the steel anchor bolts. Bend each side of this termite plate down 1 inch. It does not need to be nailed, as drawing the sill plates down with the anchor bolts will be more than sufficient to hold the shield in place (Fig. 10-3).

CONCRETE SLABS

For garages and other such structures, a concrete floor is desirable. The home garage is usually built on a slab. Keep the top of the slab above ground level to prevent seepage.

Once the outline is established, the topsoil is removed and any needed drainage and sewer lines are placed. Cover them with 6 inches of gravel, which must be tamped down well. Now lay a vapor barrier of 6 mil plastic sheeting, lapping any joints at least 4 inches. If the garage is to be heated, lay rigid foam insulation all around the outside of the slab to be laid, extending inwards at least 2 feet. Use 6 by 6-inch number 10 wire mesh to reinforce the slab. The slab must be at least 4 inches thick. Heavy duty use will require a slab 6 inches thick. Garage walls must still be supported by a footing that extends below frost depth. If the distance is great, you can save money by putting in the footing and then erecting a concrete block wall on that to hold the slab edges (Fig. 10-4).

FLOOR FRAMING

When a concrete slab is used, floor framing is seldom needed. Even if another type of flooring is to be placed on the slab, it is usually provided by using pressure-treated 2 by 4 sleepers nailed to the slab, upon which the subfloor and any finish floor is installed. For others types of bases, though, floor framing is essential. Floor framing systems start at the sill plates and consist, usually, of joists, beams, posts and subfloor. For smaller buildings (those with a span of 14 feet or less), a center beam is seldom needed. Posts probably won't be needed if the correct lumber size is used.

Platform Framing

Because outbuildings seldom reach two stories, unless they are barns (in which case pole construction is more economical by far), we'll rule out covering balloon style framing and stick with the simpler paltform framing methods. Platform framing begins with a box sill of 2-inch or thicker (nominal) lumber for

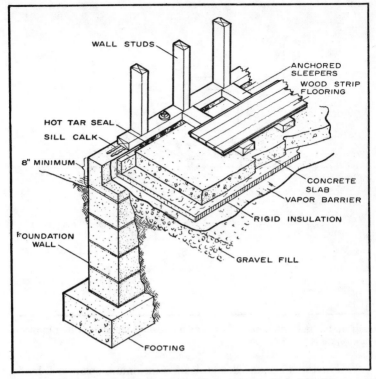

Fig. 10-4. Floor slab with independent foundation wall and footing.

the sill plate, anchored to the foundation wall bolts over a sill sealer and termite shield.

Spacing and timber size for joists will vary with the use you intend to make of the sturcture. For use with heavier workshop tools, 16-inch on-center spacing with joists one size larger than specified for the particular wood species and span for residences is a good idea. Light duty use as a studio or tool storage area requires 24-inch on-center spacing with joists meeting residential requirements for wood species and span. Extra light duty uses might allow you to get away with one size less than residences require, but I wouldn't recommend it. The floor probably won't collapse, but you'll find it tends to be shaky. The generally figured live load on a residence is 40 pounds per square foot, while even light manufacturing buildings face a loading of 125 pounds per square foot. The advisibility of going to one size larger joists when heavy tools are to be used is easily seen.

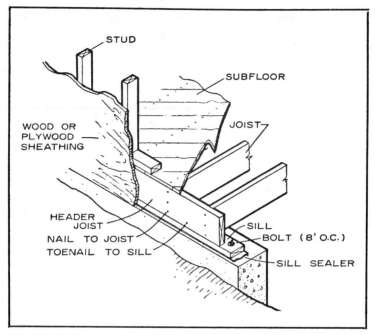

Fig. 10-5. Platform construction, again showing the anchor bolt in place, and the use of a sill sealer.

For the 40 pound design load, a maximum span for 2 by 6s, placed 12 inches on center, would be 10 feet 3 inches (for common structural lumber). Opening the on-center distance to 2 feet cuts the span to 8 feet 1 inch. Simply moving up to the same grade of 2 by 8 allows the 2-foot on-center distance over a span of 10 feet 10 inches. The same grade of 2 by 10 lumber would allow a maximum span of 13 feet 9 inches, while 2 by 12 joists would let you have a maximum width of 16 feet 7 inches (both on 24-inch centers). By moving that 2 by 12 in to 16-inch on-center spacing, you could span 19 feet. In every case I would cut the span by at least a foot and probably two to allow for dead loads (the weight of the building itself).

With platform construction, a header joist is nailed to the sills along the sides to accept the joist ends. This joist is toenailed to the sill, and the joist ends are then nailed the header joist. Use three 16 penny nails per joist end, while the header joist is toenailed to the sill with 16 penny nails spaced every 16 inches. Each joist is also toenailed to the sill, using 10 penny nails, two per joist (Figs. 10-5 and 10-6 and Table 10-1).

Plywood Subflooring

Subflooring forms a working platform over wood joists as the rest of the building is constructed, and then provides a base for any finish flooring you decide to use. The most common materials for subflooring are tongue and groove boards 5 to 8 inches wide and ¾-inch thick (nominal 1-inch lumber), or plywood from ½ to ¾-inch thick. Board subflooring is applied either diagonally across the joists (preferred) or at right angles. If board finish flooring is laid over a subfloor applied at right angles to the joists, the finish flooring must run at right angles to the subflooring. Diagonally laid subflooring gives you a choice. The finish floor boards can then be laid at either right angles to the joists or parallel to them. The subfloor is nailed to each joist with two 8 penny nails for boards up to 6 inches in width, and three 8 penny nails for boards 8 inches wide.

When using plywood for subflooring, you've got a fairly wide choice of grades. Select a type that will serve as both subfloor and finish floor. Such plywood is used when it is generally assumed the final covering will be either some form of tile or carpet. For a storage shed or workshop floor, a good finish directly on the plywood can be sufficient. No matter the grade, thickness or type plywood you use, I would recommende specifying exterior grade glue. The waterproof glue will go a long way toward making the floor last longer and only adds a little to the overall cost.

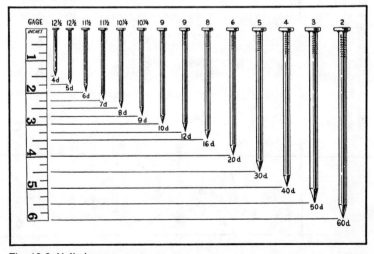

Fig. 10-6. Nail sizes.

Table 10-1. Schedule for Nailing the Framing and Sheathing of a Wood Frame House.

JOINING	NAILING METHOD	NAILS		
		NUM-BER	SIZE	PLACEMENT
Header to joist	End-nail	3	16d	
Joist to sill or girder	Toenail	2	10d or	
		3	8d	
Header and stringer joist to sill	Toenail		10d	16 in. on center
Bridging to joist	Toenail each end	2	8d	
Ledger strip to beam, 2 in. thick		3	16d	At each joist
Subfloor, boards:				
1 by 6 in. and smaller		2	8d	To each joist
1 by 8 in.		3	8d	To each joist
Subfloor, plywood:				
At edges			8d	6 in. on center
At intermediate joists			8d	8 in. on center
Subfloor (2 by 6 in., T&G) to joist or girder	Blind-nail (casing) and face-nail	2	16d	
Soleplate to stud, horizontal assembly	End-nail	2	16d	
Top plate to stud	End-nail	2	16d	At each stud
Stud to soleplate	Toenail	4	8d	
Soleplate to joist or blocking	Face-nail		16d	16 in. on center
Doubled studs	Face-nail, stagger		10d	16 in. on center
End stud of intersecting wall to exterior wall stud	Face-nail		16d	16 in. on center
Upper top plate to lower top plate	Face-nail		16d	16 in. on center
Upper top plate, laps and intersections	Face-nail	2	16d	
Continuous header, two pieces, each edge			12d	12 in. on center
Ceiling joist to top wall plates	Toenail	3	8d	
Ceiling joist laps at partition	Face-nail	4	16d	
		2	8d	

Joining	Nailing method	Number of nails	Nail size	Placement
Rafter to top plate	Toenail	5	10d	
Rafter to ceiling joist	Face-nail	3	10d	
Rafter to valley or hip rafter	Toenail	3	10d	
Ridge board to rafter	End-nail	4	8d	
Rafter to rafter through ridge board	Toenail	1	10d	
	Edge-nail			
Collar beam to rafter:				
2 in. member	Face-nail	2	12d	
1 in. member	Face-nail	3	8d	
1-in. diagonal let-in brace to each stud and plate (4 nails at top)	Face-nail	2	8d	
Built-up corner studs:				
Studs to blocking	Face-nail	2	10d	Each side
Intersecting stud to corner studs	Face-nail		16d	12 in. on center
Built-up girders and beams, three or more members	Face-nail	2	20d	32 in. on center, each side
Wall sheathing:				
1 by 8 in. or less, horizontal	Face-nail	2	8d	At each stud
1 by 6 in. or greater, diagonal	Face-nail	3	8d	At each stud
Wall sheathing, vertically applied plywood:				
3/8 in. and less thick	Face-nail		6d	6 in. edge
1/2 in. and over thick	Face-nail		8d	12 in. intermediate
Wall sheathing, vertically applied fiberboard:				
1/2-in. thick	Face-nail			1½-in. roofing nail 3 in. edge and
22/32 in. thick	Face-nail			1¾-in. roofing nail 6 in. intermediate
Roof sheathing, boards 4-, 6-, 8-in. width	Face-nail	2	8d	At each rafter
Roof sheathing, plywood:				
3/8 in. and less thick	Face-nail		6d	6 in. edge and 12 in. intermediate
1/2 in. and over thick	Face-nail		8d	

Plywood for subflooring will be marked as to maximum allowable span. For example, plywood stamped 32/16 has a maximum allowable on-center distance of 16 inches for floor use. The 32 indicates the maximum on-center diatance for use on roofs.

If the plywood is to be used for flooring that will not be covered, use tongue and groove plywood. Use glue specified by the plywood's manufacturer. Install plywood for subflooring with the grain of the outer plies running at right angles to your joists for greatest stiffness. Use 8 penny nails for ½ to ¾-inch plywood, spacing the nails 6 inches apart along the outside edges of the panel and 10 inches apart along the inside nailing spots. Don't lay plywood with dead tight joints. Some room for expansion and contraction must be allowed, so leave a gap of at least 1/16 inch at sides and ends (Fig. 10-7).

WALL FRAMING

Once the floor is framed and the subfloor is down, it's time to erect the walls. Wall framing serves two purposes. First, the exterior walls (and often interior walls) provide support for the roof, or second floor. The studding provides a surface to which interior finish materials can be nailed.

Residential construction commonly requires that 2 by 4 studs be spaced no more than 16 inches on center, with 2 by 6s allowed to go out to 24 inches on center. For workshop or other non-residence erection, your local codes may allow you to open the 2 by 4s up to 24 inches on center. Top and sole plates are of the same size material used for studs. A lot will depend on the exterior sheathing and siding. Most homes are built with a good safety margin for wind loading and downward stresses. Adding heavier than usual sheathing can increase wind load resistance to the point where many locales will allow homes to be built with 2 by 4s on 2-foot centers.

You'll probably want to use standard door sizes while constructing a wood frame outbuilding. Arrange for an interior ceiling height of 8 feet. This means buying precut studs to rough frame the wall. These studs are cut to allow a rough height of 8 feet 1½ inches, permitting a 1½-inch sole plate and 3-inch thick doubled top plates. Or cut to fit a sole plate of 1½ inches and a single top plate of 1½ inches. Using material in this manner allows the use of standard size interior materials, such as 4 by 8-foot gypsum wallboard and 4 by 8 foot wall

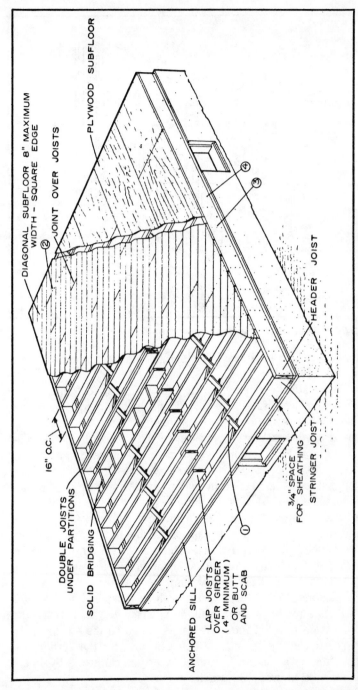

DIAGONAL SUBFLOOR 8" MAXIMUM WIDTH – SQUARE EDGE

② JOINT OVER JOISTS

PLYWOOD SUBFLOOR

16" O.C.

DOUBLE JOISTS UNDER PARTITIONS

SOLID BRIDGING

ANCHORED SILL

LAP JOISTS OVER GIRDER (4" MINIMUM) OR BUTT AND SCAB

①

3/4" SPACE FOR SHEATHING

STRINGER JOIST

HEADER JOIST

③

④

Fig. 10-7. Floor framing for platform construction, showing both plywood and board subflooring.

247

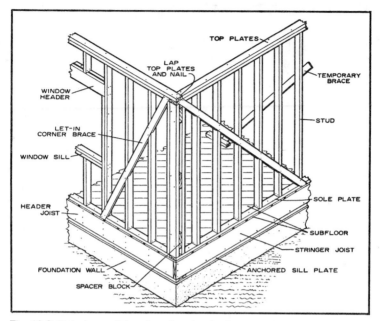

Fig. 10-8. Exterior wall framing showing let-in corner bracing.

paneling, while still leaving clearance for a finish floor and finished ceiling (Fig. 10-8).

For platform framing, you'll find it easiest to frame a wall section in place on the subfloor. Then tilt it into position where it is plumbed, braced and nailed. Construct only the length sections that you and any helpers can tilt into place. If the building is fairly large, you may have to do each wall in several sections.

Lay out the precut studs, any window and door headers, cripple studs (studs cut to fit under window sills or over doors) and window sills. Mark both the sole plate and the top plate for the stud position. Nail the studs through both the top and sole plate, using two 16 penny nails at each end. If a second top plate is needed, it is nailed last. This cuts down on the need for toenailing, a procedure that many amateur builders find tends to cause studs to skitter out of plumb. All exterior walls should have doubled top plates, even though in most building designs (with gable roofs) the end walls do not bear the load. When the wall section is tilted into place, it is nailed to the sill using 16 penny nails every 16 inches. You can now cut the corners for let-in bracing, as shown in Fig. 10-9, or you can use 2 by 4s cut and inserted between the studs.

248

Corners require an extra stud. This should be braced away from one other stud with 2 by 4 blocking. The blocking provides a nailing edge for interior finish.

FRAMING OPENINGS

As the walls are laid out on the subfloor, you'll want to ready any window or door openings that are planned. Such openings require headers which are made up of two pieces, spaced with wood lath and nailed toegether. Headers must increase in size as the opening increases. A maximum span of 40 inches would use a doubled 2 by 6 header, while a span of 5 feet would require 2 by 8s doubled. Going to 80 inches requires a header of doubled 2 by 10s, and 8-foot openings require double 2 by 12s. No matter the use of your outbuilding, it is always best to select windows and doors that are standard sizes. Cus-

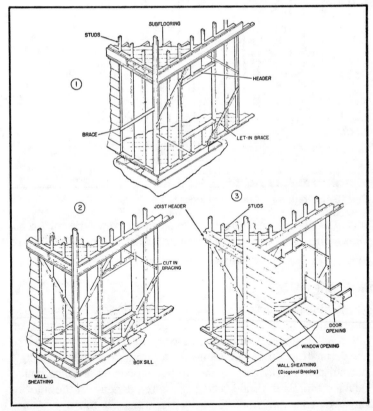

Fig. 10-9. Various types of wall bracing.

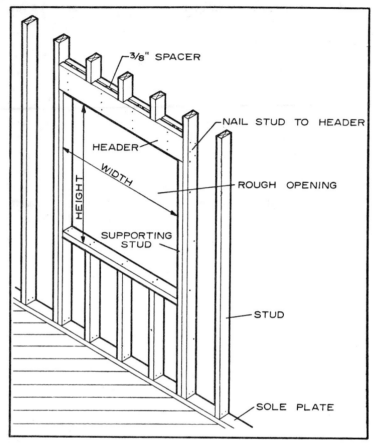

Fig. 10-10. Framing around openings.

tom windows tend to be very expensive even for residences. Check with your local supplier to see what sizes and styles are considered standard (Fig. 10-10).

INTERIOR WALLS

You may or may not need to use interior walls in your outbuilding, depending on the structure's planned use and size. There is little real difficulty in erecting interior walls, since their construction is the same as for exterior walls—a single bottom plate, studs set to the proper on-center distance, and a doubled top plate. Stud spacing will often depend on the interior finish of your walls. A 16-inch on-center wall can use ⅜-inch gypsum wallboard. If you open the on-center distance to 2 feet, you'll have to use ½-inch wallboard. Such items may not be in your local code for outbuildings, but it's still a good idea to go with them. Otherwise, you could end up with walls that drum and shake. For door openings in non-load bearing partitions, a single support stud at each side of the opening is sufficient. Use double studs if the wall is load bearing (Figs. 10-11 and 10-12).

If you use a truss roof system, none of your interior walls will be load bearing. Short spans allow you to refrain from using load bearing partitions.

CEILING JOISTS

Ceiling joists are needed if your outbuilding is to have a ceiling other than the underside of your roof. These joists are

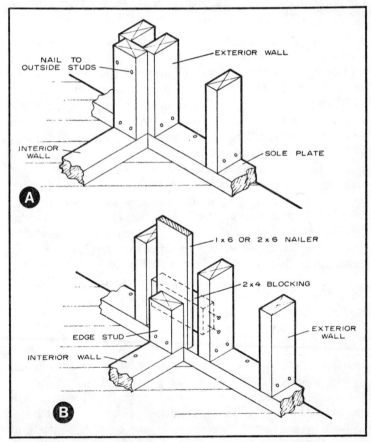

Fig. 10-11. Framing at corners.

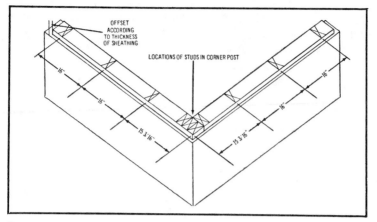

Fig. 10-12. Laying out the stud locations before erecting them simplifies and speeds the job.

placed and nailed before the rafters go up. They are usually positioned across the width of the house, as are the rafters. It is best to space any load bearing interior walls so that ceiling joists of standard (10, 12, 14, and 16 feet) sizes can be used to cover the span to these walls. Again, the size of the joists will depend on the span and the species of wood used, as well as the on-center distances. Most of the time your local building codes will give you the information you need to save a lot of figuring.

Ceiling joists also act as tension members when you erect a building with a pitched roof. In other words, they help to keep the walls from shifting outward under the pressure of the rafters. You need to make sure the ceiling joists are well nailed at the top plate, both on inner and outer walls. If two pieces are needed to cover a span, they must be securely nailed together. Either overlap them and nail directly or use cleats, as in Fig. 10-13. If your area is subject to severe windstorms, small buildings are better served by metal ties nailed to the top plate and studs (Fig. 10-14).

ROOF FRAMING

Roof framing follows the installation of ceiling joists. Usually outbuildings will use either a shed roof or a gable roof, which is the simplest form of pitched roof. If you want to use the ceiling joists as rafters, a shed roof with a slight pitch can be built by installing *cant strips* along the tops of the joists to provide some slope for rain runoff. In such cases you'll have to use larger lumber for the ceiling joists because they'll be supporting more load. Figure 10-15 shows two methods of framing a flat roof, though cant strips are not shown. Cant strips are simply timbers cut on a taper to match the desired final roof slope. Then they are nailed to the tops of the joists to provide a surface for the roof sheathing. If no side overhang is needed or desired, you would simply continue the rafters out to the edge of the structure instead of using lookout rafters as in Fig. 10-15.

Gable roofs tend to be a bit more coplex than shed roofs. Once the first pair of rafters is cut, all others can be cut to the same pattern. To save time and the figuring needed for getting the first set of rafters right, you can use trusses. Trusses eliminate attic space, but provide a reasonably low cost and fast method for the amateur to erect a perfect roof with no cutting needed.

Rafters are not nailed in place until the ceiling joists are in. But trusses require no load bearing interior partitions and may be placed as soon as the wall framing is up. Figures 10-16A through 10-16C shows three different styles of roof trusses popular for small building construction. You can, of course, go ahead and make your own trusses. Figure 10-17 shows the typical construction of a truss for a 26-foot span with a rise of 4 in 12. For such a truss, you would use 2 by 4 members braced with guussets made from ½-inch plywood. The on-center distance could be 2 feet. This would provide for a total roof load of 40 pounds per square foot, which would be sufficient almost

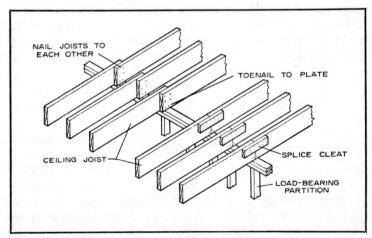

Fig. 10-13. Ceiling joists over a bearing partition.

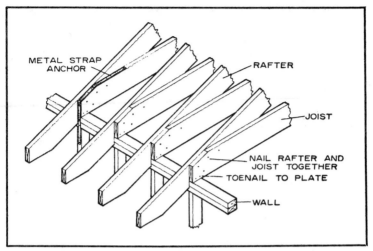

Fig. 10-14. Ceiling joist connections on a gable roof.

everywhere except in those states where snow loading is exceptionally heavy.

RAFTER LAYOUT

When trusses are not used, you'll need to know how to determine rafter length and top and bottom angles for your building. The first job is to check the span of the building, using a good quality tape measure. Measure to the outside of the top plates. Now take half that distance, so that on a 20-foot span you have a run of 10 feet. If your rise is to be 8 inches per foot, check your carpenter's square. Mark the tongue at 8 inches and the blade at a foot. Measure the diagonal formed which is 14 5/12 inches in this case. Since the run of the rafter is 10 feet, you would multiply the line length of 14 5/12 inches by 10 and get a total rafter length of 12 1/6 inches. Normally, at least a foot of overhang is added, so you would need 14-foot long members for your rafters.

Now that you have the length and the material, select as straight a piece as possible and lay it across a couple of sawhorses. If there is any crown or bow to the rafter, make sure it is kept to the top, or what will be the top, of the rafter. Take the carpenter's square, marked on the tongue at 8 inches and on the blade at 1 foot, and place the square as shown in Fig. 10-18. This will give you the plumb cut for the ridge board. Finally, you can move on down the rafter and measure off the 12

1/6 inches. Hold the square with the 8-inch mark directly over the rafter length mark. This gives the overhang cut line. Take the measurement of your top plate. Set the square on the 8 and 12 marks and measure off the width of the notch needed from the heel of the square. Then mark the notch and cut.

For those who don't trust their measuring skills quite so much, another method of marking off the rafter is available. It is probably a bit simpler as you need only be sure of accuracy as

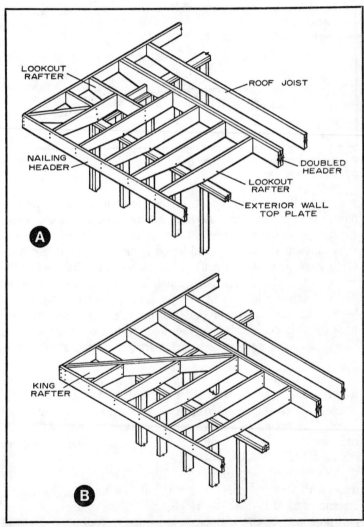

Fig. 10-15. Two methods of framing the overhang on a flat or shed roof.

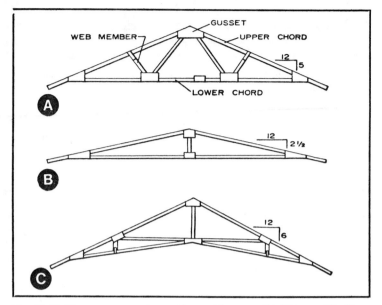

Fig. 10-16. Three common types of wood trusses. (A). W-type. (B). King-post. (C). Scissors.

you step off down the length of the rafter material. From this point, you "stake" or measure the material using the same number of stakes as there are feet in the run. If there is a left-over amount (as there would be on a run of, say, 10 feet 4 inches), you measure and mark 4 inches out from the heel of the square when the square is in the last stake, or step, position. Then measure using the mark as a setting point for the rafter's heel. With the square in the final position, you now measure back the distance needed for the notch to fit over the top plate. Then move the square back and mark the angle, as shown in Fig. 10-19.

Don't forget to remove from each rafter half the amount of material that is supplied by the ridge board. In other words, if a nominal 1 inch and actual ¾-inch ridge board is used, you would make a cut taking off ⅜ inch on each rafter top end. After cutting two rafters, check for fit. If all is well, the remaining rafters can be cut (Fig. 10-20).

You'll need help erecting rafters. The job is most easily done with three people or more, but can be done by two if bracing is used to hold the ridge board in its approximate correct position. Rafters are placed in pairs. Start with one end and

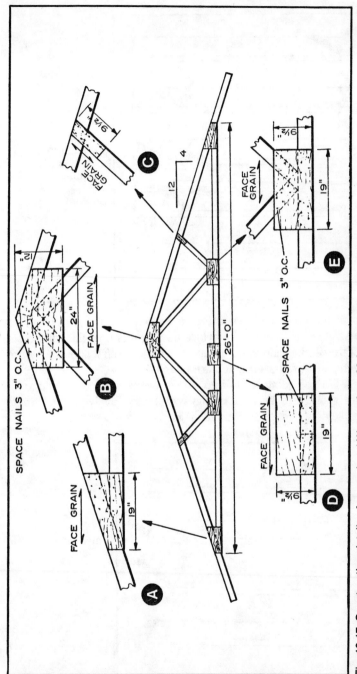

Fig. 10-17. Construction details of a wood end W truss for a 26-foot span.

257

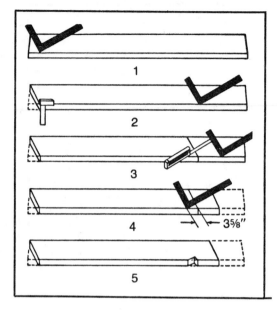

Fig. 10-18. Scale method of rafter layout.

go to the other end. Then fill in the centers. Use lines and Xs to set the rafter locations before placing them, as this makes accurate placing and nailing easier. Check carefully to make sure each pair of rafters erected are plumb.

I'm not going to go into valley framing, hip roofs and gambrel roofs here because of the complexity involved. Such roofs are seldom needed for outbuildings.

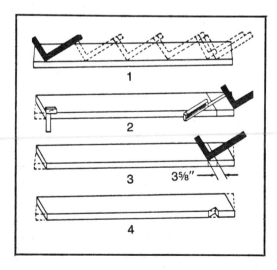

Fig. 10-19. Step off method of rafter layout.

ROOF SHEATHING

There is a variety of *roof sheathing* available. The proper selection is important to the overall structural rigidity of your building. Sheathing goes on directly over the rafters and then is covered with some type of roofing. You can use butt joined sheathing board, tongue and groove sheathing board or plywood. Nominal 1-inch lumber is needed for the first two. Plywood in the proper thickness as specified by the manufacturer for on-center spacing is used for the latter. A standard sheathing grade of plywood will be marked for the maximum on-center distance. Plywood marked 16/32 will be used for 16-inch on-center sub-flooring, or 32-inch on-center roofing sheathing.

Board sheathing can take two forms: closed sheathing and open sheathing (Fig. 10-21). In either case, all joints are made over the centers of rafters. Generally open sheathing is prefer-

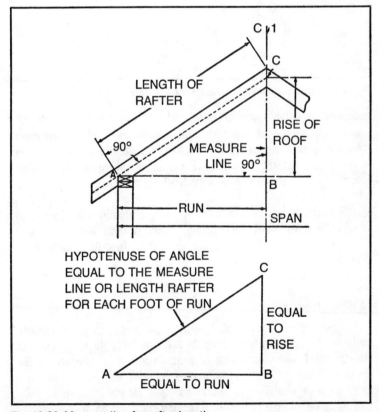

Fig. 10-20. Measure line for rafter length.

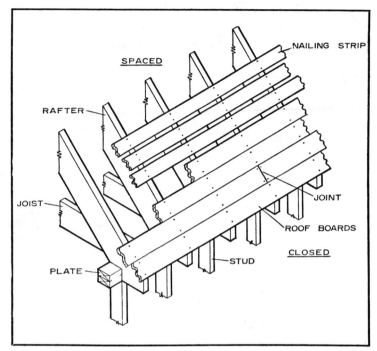

Fig. 10-21. Closed and open board roof sheathing.

red with wood shingle roofing and corrugated metal roofing. Closed sheating is better with asphalt shingles. Eight penny nails are used to nail sheathing. Board sheathing should be long enough to span at least three rafters. Plywood sheating is laid so that boards directly above or below each other do not break on the same rafter.

Sheathing around chimney openings, should you decide to provide heat for your outbuilding, stops ¾ inch from the chimney. No rafter should be closer than 2 inches to the chimney (Fig. 10-22).

LAYING SHINGLES

Roofing materials will depend in part on your overall building design. Asphalt shingles are not suitable for shed roof construction, so some form of metal roofing is needed. Gable roofs will accept (assuming a rise of at least 3½ inches in a foot) asphalt shingles or metal roofing. Heavy roofing in rolls can also be used. The roll roofing tends to be less durable, though it is a fair amount cheaper and faster to put down (Table 10-2).

The first chore in laying shingles, or roll roofing, on a gable roof is the laying of 15-pound roofing felt. Use a 4 inch overlap when laying this, with the overlap coming from the top piece so water will drain down (Fig. 10-23). The first or starter course of shingles can be laid. This first course is inverted—that is, the open or serrated edges point to the ridge board. Each course of shingles is started with a chalk line snapped across the roof, allowing the correct amount of the shingle to be exposed to the weather. The manufacturer provides this information, but for most asphalt shingles the distance is 5 inches. Use 1-inch galvanized nails and place two nails to each tab, making six per shingle. The nails should be placed far enough up towards the butt (solid) edge of the shingle to make sure it is covered by the next course, but not so far up as to allow too much flapping around. Six to 6½ inches down from the butt edge is usually fine. Such nailing is called *blind nailing*.

After the starter course is laid, it is covered with the first course and the shingling is continued on up to the ridge. Shingle

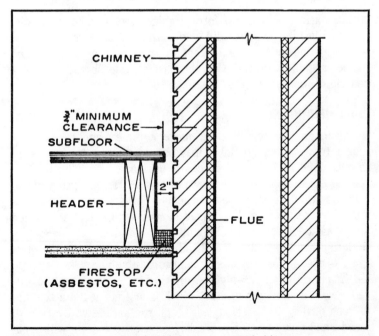

Fig. 10-22. Wood construction clearances around chimneys.

Table 10-2. Minimum Roof Slope and Approximate Weight of Various Roof Coverings.

TYPE OF ROOFING	MINIMUM RISE PER FOOT RUN WITH ORDINARY LAP	APPROXIMATE WEIGHT PER SQUARE
	Inches	Pounds
Aluminum...	4	30
Asbestos shingle:		
American multiple	5	300
American ranch..................................	5	260
Asbestos, corrugated	3	300
Asphalt shingle:		
Lockdown...	4	290
3-tab...	4	210
Built-up roofing...................................	½·	600
Canvas (8 to 12 ounce)	½·	25
Galvanized steel:		
Corrugated..	4	100
V-crimp...	2½·	100
Roll roofing:		
Regular (2- to 4-inch lap)	4	100
Selvage edge (17- to 19-inch lap)........	1	140
Slate	6	800
Tin:		
Standing seam..................................	3	75
Flat seam..	½·	75
Wood shingles..	6	200

The different types of roofing vary in weight per square according to the weight or thickness of the roofing material itself.

squares are now used to form a ridge line, as shown in Fig. 10-24. Flashing can be laid under this to provide long-term water tightness. All but the last ridge shingle section is blind nailed, and the nail heads on this last shingle should receive a coating of roofing tar (Fig. 10-25).

BUILT-UP ROOFING

Built-up roofing is a bit more complex and takes a bit longer to lay properly. But it may be essential if you don't want to use metal roofing on a shed roof design. Essentially, built-up roofing consists of several layers of asphalt rag felt paper set in a hot material, usually melted asphalt. A last layer of binder is spread over the top and sprinkled heavily with gravel or crushed stone as a last step. While built-up roofing for residences is preferably laid over doubled sheathing, a good, tight single sheathing is usually more than sufficient for outbuildings.

Properly done, a 5-ply built-up roof will last a couple of decades with a minimum of care, which is just about the actual life of good quality asphalt shingles. Metal roofing life will vary with weight and metal type. Eventually, galvanized roofing will begin to rust as hail, tree branches and other things wear the surface coating off. At this point, a good steel brushing and a high quality primer can be used, after any needed patching, as a base for paint to extend the lifetime of the roof.

Nailing points should be predetermined for built-up roofing so that the nails in successive layers will not strike nails in lower layers. Start at the eaves and unroll the building paper or

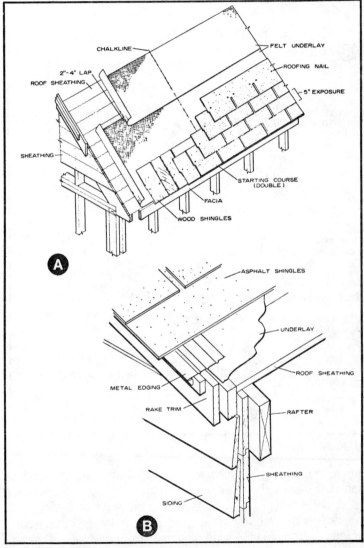

Fig. 10-23. Application of asphalt shingles.

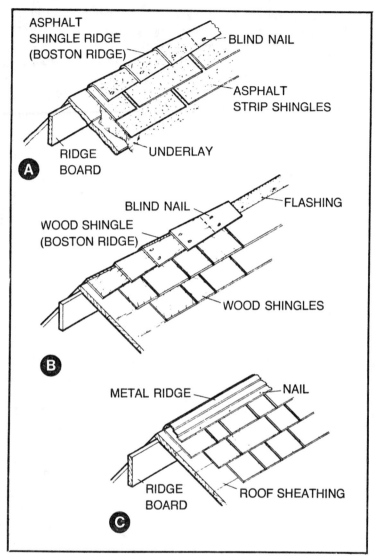

Fig. 10-24. Ridge finishes.

roofing felt, using a 16-inch wide strip saturated in the binder asphalt. Space nails about 1 inch from the back edge of the paper, about a foot apart. Start with a full strip of roofing felt under the saturated 16-inch strip to add durability (Fig. 10-26). Now lay a 32-inch wide strip over the two already down, using no binder between the layers. Use the same nailing pattern as

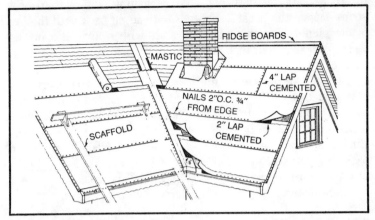

Fig. 10-25. Roll roofing installation.

you did for the 16-inch strip. The next full width strip is laid with its bottom edge 14 inches from the top edge of the first two strips laid, giving a 2-inch overlap. Continue laying full width strips with the 2-inch overlap unti the roof is covered. Start the next layer 14 inches back, providing the same overlap and exposure. These two first layers are laid dry.

For the next three plies or layers, you'll need to rent a pressure kettle to keep the asphalt at the correct temperature, usually about 400 degrees. Use a mop to apply the asphalt to

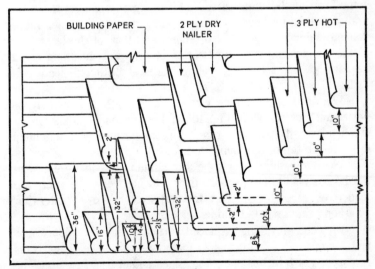

Fig. 10-26. Laying a 5-ply built-up roof.

the roof. Start the third ply, after asphalting, with a third of a strip—lapped the usual 2 inches. Continue on back until the ply is completed. For the final ply, start with a strip 8⅔ inches wide. Provide, from that point, for a 10-inch exposure on each succeeding strip.

If your kettle won't quite maintain the correct temperature, it is best to go to a slightly cooler setting instead of one that's too hot. Binder which is heated too much can burn the roofing paper or form a very thin layer which will eventually crack. Binder that is slightly too cool for easy spreading will wear okay, but you will use more material than you would need otherwise. It's better to have a slightly too heavy buildup and use a bit more material to get greater durability.

WALL SHEATHING

Not all outbuildings will use wall sheathing. For those that do, the choices today are wider than ever. Many forms of plywood are available for use as a combination of final siding and sheathing. There is, standard nominal 6 inch by 1 inch sheathing board, as well as structural insulating board and gypsum board. Again, requirements differ from area to area. Most barns don't require sheathing. The board siding is usually applied directly to the particular form of framework. The same holds true for sheds used for tool and material storage. For buildings to be heated and used as workshops or studios, I would recommend some form of sheathing to add a bit of rigidity and cut down on drafts. For barns used to house foaling mares and other more delicate animals, I would also recommend that, at the least, one of the tongue and groove varieties of plywood sheathing could be used for barns housing foaling mares and other delicate animals.

Wood board sheathing can be placed either diagonally or horizontally as shown in Fig. 10-27. The diagonal installation increases overall structural rigidity, particularly if it is extended down to the top of the foundation.

Wood sheathing should be a minimum of ¾-inch thick, and the boards should be no more than 6 inches wide. All joints are made over studs for standard framing styles, or over nailers for post and beam or pole building styles.

Plywood sheathing is placed vertically in most applications and goes up much more rapidly than does board sheathing. Nails are spaced at 3 inches along the edges and at 6 inches along the

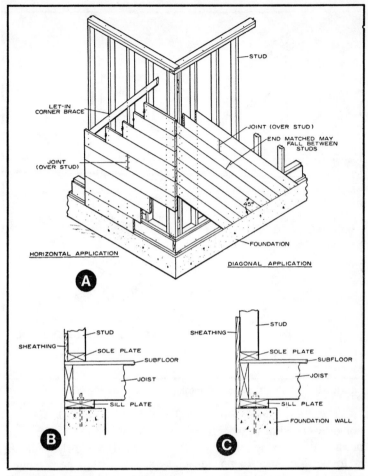

Fig. 10-27. Applying wood sheathing.

interior lines. Use the appropriate thickness of plywood for your on-center stud spacing. Check the figures stamped on the plywood by the manufacturer to determine if you have the correct spacing. Six penny nails are used for ⅜-inch sheathing, and 8 penny nails for ½-inch or more sheathing. If corner bracing was used to brace the house framing, you can open up nail spacing to 6 inches along the outside edges and a foot along interior nailing areas. Horizontal plywood applications can also be used, but they don't aid rigidity as does vertical application. Edge and end spacing should be from 1/16 to ⅛ inch to allow for expansion and contraction of the plywood (Fig. 10-28).

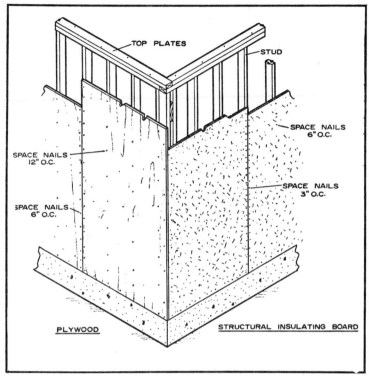

Fig. 10-28. Applying plywood sheathing.

Sheathing paper is often called building paper. *Do not* use polyethylene film or sheeting here. The sheathing paper must be able to "breathe" if the moisture trapped inside the building is to get out. Use 15 pound builder's felt, with a permeability rating of no less than 6. For plywood sheathing, the paper is needed only around window and door openings to cut down air infiltration.

SIDING

Companies such as Georgia Pacific make a wide variety of plywood sidings to give the effect of board and batten, board on board and about any other style you can imagine. While a bit more expensive than actual board on batten construction, such large sheets of siding go up much more rapidly.

Many outbuildings are unsheathed and have a corrugated or plain metal siding. These types of siding are available in a wide variety of sizes and colors. Sears has ribbed aluminim building panels in natural or white, suitable for use on roofs or as siding. Width is 48 inches, but lengths go from 8 to 16 feet. Fiber glass panels come in 40-inch widths and in 8, 10 and 12-foot lengths, either white or green. These fiber glass panels are translucent. Sears offers the panels in a variety of weights and colors. You can use lighter weights for siding jobs and save some money. When installing any type of siding, use nails of the appropriate type (nails for fiber glass and metal sidings will have built-in washers) with a screw or ring shank for best holding power.

Natural wood sidings offer some great savings over plywood, especially if there are sawmills cutting timber species nearby. With inflation, it is really impossible to predict the possible savings. If you decide to opt for local material, make sure it has at least three months of reasonably dry weather in which to season. Keep it stacked correctly, in single board layers with cleats between each layer, to allow for good air circulation. Weight the top layer with concrete block or some other heavu material to help cut down on warping. Cover the top only with plastic sheeting.

Once your siding is on, apply a good penta base preservative to aid long life. Paint or stain if desired.

DOORS

In most cases today doors will be of the pre-hung style, with hinges and other hardware already in place. The job is then simply one of being accurate and getting things plumb. Assuming the rough framed opening is plumb, the door should fit in place with a minimum of fuss. Use just a few pieces of shim stock to fill any open spaces between the jamb and the rough framing. Plumb the door in two directions and get it accurate. You don't want a door that sticks.

PANELING AND WALLBOARD

The insides of some outbuildings will require finishing of some sort, either paneling or wallboard. Both processes are reasonably straightforward. In my opinion, the installation of paneling is an easier job since there are no nail dimples or seams to be filled afterwards. Use as a light a saw as possible to cut the paneling, and make the cuts accurate. Use color-matched nails or glue to the studs. That's it.

Table 10-3. Gypsum Board Thickness.

| INSTALLED LONG DIRECTION OF SHEET | MINIMUM THICKNESS | MAXIMUM SPACING OF SUPPORTS (ON CENTER) | |
		WALLS	CEILINGS
	In.	In.	In.
Parallel to	3/8	16	—
framing members	1/2·	24	16
	5/8	24	16
	3/8	16	16
Right angles to	1/2·	24	24
framing members	5/8	24	24

Sheetrock is a brand name. The material is properly called gypsum wallboard. It is a gypsum filler faced with light-colored paper and is now used almost entirely in place of plaster walls. Gypsum wallboard comes in sheets 4 feet wide, with lengths ranging from 8 to 16 feet. Edges are tapered to allow for taping and joint filling. For outbuilding use, a 3/8-inch thick wallboard is usually sufficient. If you ever want to convert your studio or workshop to a residence, check your local codes.

Many require 1/2-inch thickness or two 3/8-inch thicknesses of wallboard. Table 10-3 shows required wallboard thicknesses for various on-center spacings, both for walls and ceilings. These requirements are for residences and may well not be required by the code in your area. Use 5 penny ring shank nails for 1/2-inch wallboard and 4 penny ring shanks for 3/8-inch wallboard. Space nails at 6-inch intervals for wall installations and 5 inches for ceilings. Leave 3/8-inch spaces at all joints.

Whenever possible, gypsum wallboard is best installed horizontally, with any joints kept under windows or other openings. This reduces the number of vertical joints which, if not perfectly taped, are more visible than horizontal joints (Fig. 10-29A through 10-29C). When installing the wallboard, drive the nails so that the hammer head forms a slight dimple around the nail without breaking the paper surface. This provides a depression so that the nailhead is covered with joint compound. Premixed joint compound is your best bet for a smooth job. (Figs. 10-30A through 10-30D).

To tape joints, spread a fairly thick layer of joint compound and then press the tape into the compound with the joint knife.

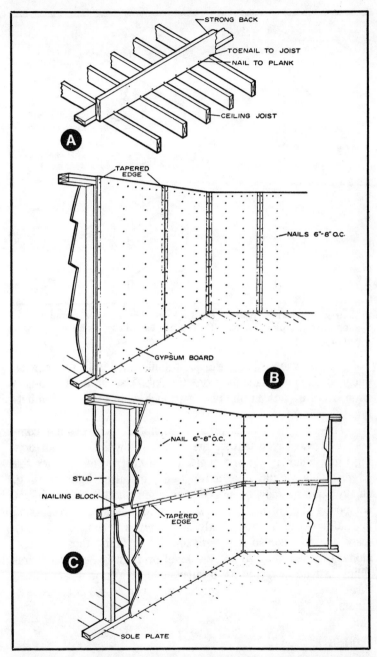

Fig. 10-29. Applying gypsum wallboard. (A) Strong back. (B). Vertical application. (C). Horizontal application.

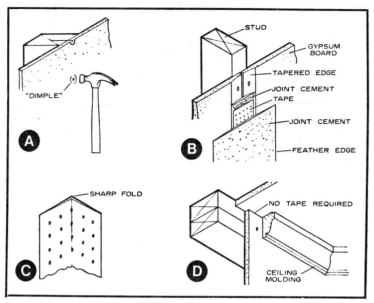

Fig. 10-30. Finishing dry wall. (A). Nail set with crowned hammer. (B). Cementing and taping the joint. (C). Taping at inside corners. (D). Alternate finish at ceiling.

You'll note when buying the joint knife that it is not a wide putty knife. The blade is more flexible. Use at least a 6-inch wide knife. If you can find one, a 10-inch wide knife is even better.

The tape is now covered with another coat of joint compound on which you feather the edges. Allow 24 hours for drying time. Sand lightly and apply a final coat; again, feather the edges. This should give a finished joint after light sanding. Heavy sanding may be needed, or even another coat of joint compound. To fill nail dimples, press in one coat of compound and allow to dry. Sand lightly and apply a second coat. Don't make the mistake so many starters do in filling these depressions. Use the knife to apply a level or slightly depressed first filling instead of trying for a huge buildup. You may even need three coats this way, but it cuts way down on sanding requirements.

FLOORS

Finish flooring materials of every conceivable kind are available. You need to consider your requirements as far as appearance, durability and cost. Then select the material that best matches your needs. In almost every case wood strip or parquet flooring will not be needed in an outbuilding. It tends to run costs up and requires care and skill in installation, even with today's prefinished varieties.

If a subfloor is not good enough as finish floor, you may want some form of plastic (vinyl) tile or resilient flooring. In all such cases, follow the manufacturer's directions as to the use of adhesive and actual installation precedures. Make sure that the joints of the flooring do not match up with the joints of the subflooring.

INTERIOR TRIM

Interior trim of many kinds is used in residences, from casing around door frames to base and ceiling moldings in different styles. The need for such trim in outbuildings is generally not heavy. For gypsum wallboard walls and ceilings, I would

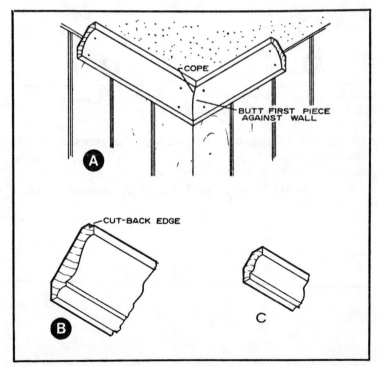

Fig. 10-31. Molding styles. (A) Installation. (B). Crown molding. (C). Small crown molding.

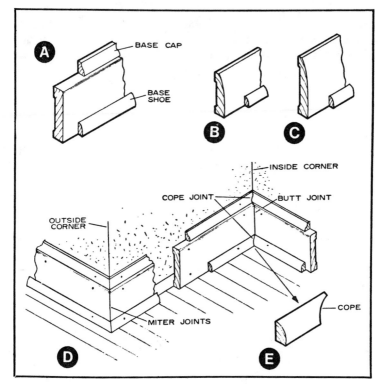

Fig. 10-32. Floor moldings. (A) Square-edge base. (B). Narrow ranch base. (C) Wide ranch base. (D). Installation. (E). Cope.

recommend installing a ceiling molding. This cuts the need for taping an inside corner joint, which is quite a job. Simple baseboards with a quarter round base shoe can also be used. See Figs. 10-31A through 10-31C and 10-32A through 10-32E.

Chapter 11
Garages

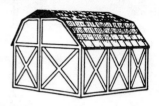

Garages can be added on to older homes as attached sections of the house, or they can be erected freestanding. In either case, it is a good idea to keep siding styles and overall design as close as is possible to that of the original structure. In other words, if you have a brick veneer house, select the same type of brick for the garage. If the house is white clapboard, then use white clapboard on the garage.

Carports are easier. Simply place posts at correct intervals, and add some type of roof. I've seen some quite attractive and simple carports put up using a nearly flat roof of fiber glass panel and decorative black metal posts.

Garage and carport selection may be limited by your site. Again, talk with the local building inspector to find out what the rules are. Driveway location and size may also be a factor.

DETACHED AND ATTACHED GARAGES

Here are my own prejudices as far as garages go. I do not care for attached garages and I hate the so-called basement garage. Actually, these are less prejudices than reactions to very real possibilities of problems and dangers inherent in such designs.

Garages are repositories for all sorts of things. People have been reminded frequently not to store such flammable items as gasoline and kerosene indoors. Yet almost every garage I've ever been in has gasoline cans, old paint cans and other fire hazards. A basement garage can prove a source of danger in other ways. The biggest problem is from the idling engine, producing carbon monoxide. It is possible for someone to start a vehicle in such a garage and then go back into the house to pick up a forgotten item. Once back in the house, it becomes simple to get interrupted and forget the running car, which may pro-

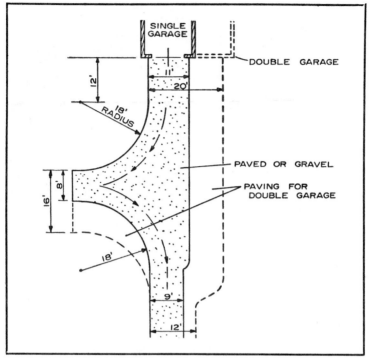

Fig. 11-1. Driveway with turnaround.

duce a fair amount of poisonous gas to seep upwards into the home.

A detached garage offers some safety advantages. It can offer some disadvantages as well, since the cost for four walls is always going to be higher than the cost for three. Still, I believe the extra cost is justified.

DRIVEWAYS

We're pretty much covered brick driveway construction, but there are other methods to look at. A too narrow driveway or a driveway with too sharp a radius in a curve can be more of a problem than no driveway at all. The need for enough room to safely enter and leave the driveway isn't going to decrease in the immediate future.

Grade

The grade of the driveway is important in all cases, but especially so if a garage or carport is planned. If the drive is

276

one with a grade of more than 7 percent, it almost has to be paved to keep things from washing away. That's a rise of 7 feet in 100 feet. If at all possible, driveways that must be steep should have at least 12 feet of level or near level area in front of the garage. Whenever possible, any grade should slope away from the garage site to keep water from draining into the garage.

Turnaround

Figure 11-1 shows a driveway turnaround that can be adapted to either single or double garages. Curve radius is 18 feet on the turnaround. The turnaround itself is 8 feet wide (for a single car garage), with a 9-foot wide drive. Nine feet is about the minimum for today's cars, though smaller vehicles might get away with 8 feet. If you plan to use the drive as a walkway as well, add a couple of feet for safety. As Figs. 11-2A and 11-2B show, the minimum curve radius at the curb (where the driveway enters the street) should be 5 feet.

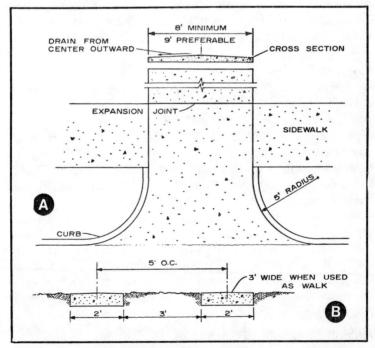

Fig. 11-2. Driveway details. (A). Single slab driveway. (B). Ribbon type driveway.

Laying Concrete Driveways

Poured concrete driveways are popular in many areas. With the ever increasing price of concrete, though, a great many asphalt drives are being constructed. Asphalt, though, is not really a suitable material for the do-it-yourselfer. Concrete is.

To lay concrete for a drive, you decide on slab thickness. Four inches will usually be sufficient if your base is good. If the base is poor or you need a heavier duty drive, go to 6 inches and lay 2 to 4 inches of gravel first. Use a 2 by 6 for a side form and you'll end up with a 5½-inch concrete drive. It is probably best to get the concrete in the form of transit mix. Then do the work of screeding yourself. Use steel mesh, 4 by 4-inch or 6 by 6-inch, for reinforcement if you want great strength.

Lay out the drive using mason's cord on stakes. Then cut away the earth and provide any needed gravel base. Set the forms in the cutout area in the design and size you prefer.

The concrete you order from the transit mix company should be what is called a five bag mix. You can estimate the cubic yards required by determining the drive's square footage. Then multiply by slab depth in feet. Four inches would ⅓ foot, while 6 inches would be ½. Divide the result by 27 to get the cubic yards.

Screeding

Screeding is done once the concrete is poured in the forms. A board long enough to fit all the way across the forms is used and moved from one end to the other with a sawing motion to settle the concrete and give a moderately rough, level finish. If you wish a smoother surface on your drive, you can now take a float (essentially a wood board with a handle on top) and work the concrete to get that effect. In very few cases would you want a smooth enough surface on a driveway slab to use a steel float. In either case, the concrete should not be worked too much while it is still extremely plastic. This tends to bring a thin, watery layer to the top which is weak and will break up easily. If you want a really non-skid surface, take a stiff straw broom and sweep the surface in a swirl pattern after troweling but before the concrete sets up all the way. Keep the slab damp and unused for at least seven days after it is laid.

ANCHORING GARAGES TO SLABS

If a garage is anticipated, lay the slab for it while the driveway is being poured. You may want to steel trowel the garage floor to aid in getting it clean after it dries. Steel troweling is not done before the moisture sheen of freshly poured concrete disappears.

Garages are anchored to their slabs in much the same manner as frame buildings. For a long lasting garage, you'll need a footing at the slab edges. That footing should be designed below frost depth, twice as wide as any needed foundation wall and at least as deep as the wall is wide.

Basic garage construction is just like any other form of platform framing—sill plates, studs, bottom or sole plates, top plates doubled, usually 16 or 24 inches on center.

SIZE AND DESIGN CONSIDERATIONS

A few considerations as to size and overall design are needed, though, that differ from general frame construction. Give some thought to the total use of any garage, as that will affect overall size a great deal. If all you really want is room to park one car with a few rafters to store items such as rakes, then the minimum would probably be a garage about 21 feet long and 11 wide. This size provides door opening room, as well as space to allow you to walk around all but the largest cars. For a workbench, or tool storage, at the end of a single car garage, allow 6 feet more length and at least 2 feet more width. A double garage should be no less than 22 feet wide.

Use anchor bolts from the foundation through the sill plate at the normal 8-foot intervals, but add one extra anchor bolt at each side of the garage door. Framing is not a huge problem for garage doors since they are usually put in non-load bearing end walls. You can use smaller headers than would otherwise be needed for such a wide space.

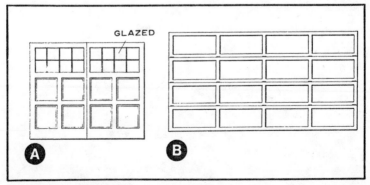

Fig. 11-3. Garage doors.

Garage doors of the overhead type are normally 9 feet wide (single) and 7 feet high, with double doors 16 feet wide. You'll need about a foot of clearance over most of these overhead doors to allow the installation of the operating mechanism, as well as room for the door to swing up into its open position (Fig. 11-3).

For the header over the door, use at least two 2 by 10s over an 8 to 9-foot span. For wider openings, it's probably best to go to a steel beam of a size recommended by your local building codes. Much will depend on your snow load and the other local variables.

Interior garage finish is something most people don't bother with. If needed, some form of gypsum wallboard or a relatively inexpensive paneling can be used. If you do build an attached garage, check local fire codes as you may need a fire-resistant material on the interior of the attached wall.

You may wish to add windows to a garage. While the ventilation is rather nice when you're working out there in the summer, the visibility windows provide to those less honest than ourselves could pose a problem. Of course, shades or blinds could alleviate that problem. A nice touch for detached garages is a small (30 inches or so wide) exterior door placed on the side facing the house. That saves having to lift the overhead door when you have only come out to pick up a garden hose or a screwdriver.

Chapter 12
Porches

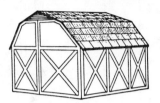

An attached porch may seem like a repetition of a deck, but porches are roofed structures which may or may not be enclosed. They tend to serve slightly different purposes than a deck. Porches can serve as greeting places for people and as places for relaxation. Enclosed or screened porches can provide fresh air on warm days while keeping insects at bay.

A porch, like any other permanent structure, begins with footings and foundations. The porches may be constructed over a crawl space, on piers, on a slab (which may also serve as the porch floor), or the basement may be enlarged and a full section added. If a crawl space is used, place several removable sections around the porch to provide access. If the crawl space is fully enclosed, it must be vented to allow air flow and cut down on moisture buildup.

To add a porch, you first must check, local building codes. Then draw up a design that you find acceptable. Plan the intersections with the existing roof and with the walls so that your porch doesn't end in the middle of windows, doors, skylights or chimneys.

The siding must be removed, as the shingles for the roof junction so that you can get at the rafters, ceiling joists and studs. Even with a gable roof, things may now become a bit more complex. You will have to lay a series of valley rafters to construct the porch roof and connect it to the old. We'll return to porch roof framing in a bit, but starting from the ground up is a better idea.

FLOOR FRAMING

Once the footing and foundation types are decided upon, they are constructed as already described earlier. Generally, porch floors are framed just like any others, but are not con-

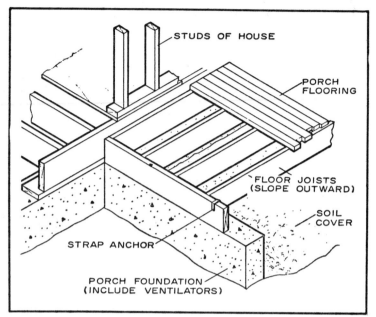

Fig. 12-1. Basic porch foundation and flooring.

structed quite level. Some pitch must be allowed for drainage if the porch is not to be fully enclosed. Usually ¼-inch per foot pitch will be sufficient (Fig. 12-1).

Floor framing for porches constructed over earth must be at least 18 inches above the ground and should be done only with pressure-treated wood. Even then, it is a good idea to use 4 mil poly sheeting as a ground cover to cut down on the rise of moisture.

Once the floor is framed, you'll want to put in any corner and line posts needed to hold up the roof. The size and location of these posts depends on several factors, including the overall porch size as well as the size of rafters and ceiling joists (if any). Most of the time, posts on 8-foot centers will be sufficient, with 2 by 8 ceiling joists, 2 by 6 rafters and a roof pitch of at least 4 in 12, if the joists and rafters are on 16-inch centers. Enclosed porches use the same framing as do the exterior walls of the house, with openings for windows and doors framed in the same mannner.

The porch floor, if not a concrete slab, will be some form of wood. It is usually a good idea to select porch floor materials for good durability, and to go with pressure-treated woods where moisture conditions may be bad and the porch is to be open. Oddly enough, for open porch use, you'll probably find that the softwoods such as southern pine serve better than oak or hard maple, since their durability when exposed to weather is better.

COLUMNS

There are many ways to form columns for porches and about as many styles. In fact, it is possible to buy prefabricated columns of various metals and woods in many designs, from formal to more casual. In most cases it is probably simpler and cheaper to go with one you make yourself. I'll be ripping the old porch off this house in a few weeks and will use Kopper's Wolmanized lumber in 4 by 6 size for the posts. These will continue on into the ground to serve as piers as well as porch roof supports. The original porch was simply built on three piers of concrete block, with the concrete block just standing on the ground.

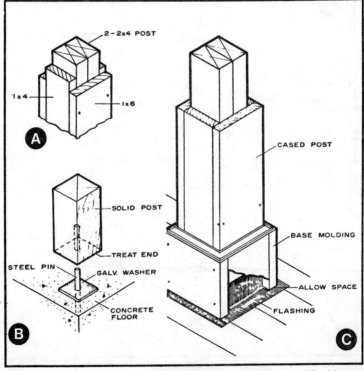

Fig. 12-2. Post details for a cased post, a solid post and base flashing.

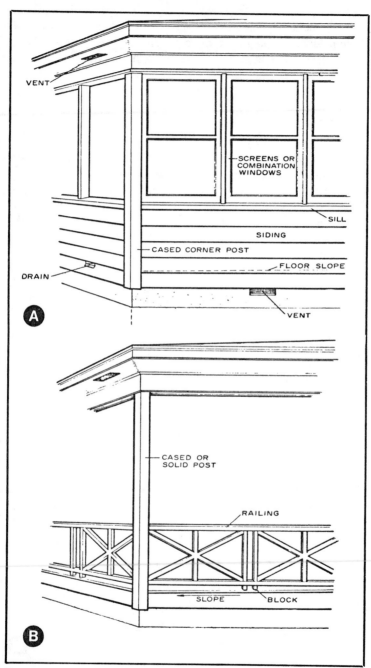

Fig. 12-3. Closed and open balustrades.

Porch posts or columns can also be formed by nailing 2 by 4s together and then casing them with 1 by 4 number 1 pine to give a finished look. Actually, you'll also need a couple of 1 by 6s and a couple of 1 by 4s, or there will be gaps. Solid posts, treated or untreated, are readily available in 4 by 4 and 4 by 6 or larger sizes. Cased posts should be placed so that moisture cannot penetrate the bottoms on open porches (Fig. 12-2).

BALUSTRADES

Bannisters on stairs are called *balustrades* on porches. Used for both decoration and to keep people from stepping off the

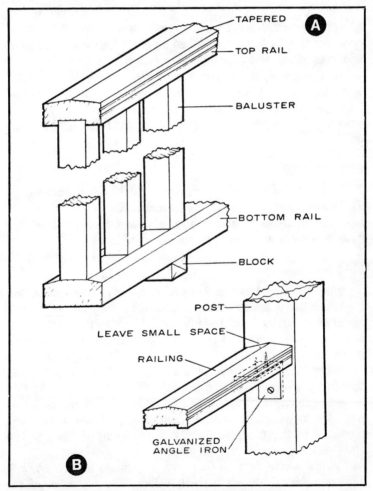

Fig. 12-4. Railing details, including balustrade assembly.

wrong side (s) of the porch, balustrades can also be bought cut to the more formal styles. Make sure the uprights, if not posts or columns, are firmly attached to the porch floor or framing. Otherwise, the railings are pretty much useless. If the wood of the uprights must be in contact with concrete or the ground, make sure to use pressure-treated lumber (Figs. 12-3 and 12-4).

ENCLOSED PORCHES

Enclosed porches can provide sunroom areas, or a screened sitting area. In most cases, the walls are framed exactly as are the interior and exterior house walls, allowing normal spacing for windows and doors. Should a large window space be desired, allow for this by bringing doubled studs up between each set of windows and by using larger header material.

If the space for the porch windows is really large and the porch roof steeply peaked to carry heavy snow loads, you may find yourself having to go with steel lintels or headers. For most porch construction, the unit size is not large enough to require steel lintels.

ROOF FRAMING

Basically, two types of roofing will be used on porches. The simplest, as with other forms of building, is the shed roof. It is the simplest to attach and the easiest to build. If a shed roof fits your home style, I would say go with it.

Shed Roof

As Fig. 12-5 shows, a *shed roof* is easily attached just below the top plate of the exterior wall. Remove the top row (or two rows) of siding, and nail the joists or rafters to the house studs as shown. Use a doubled beam of appropriate size at the outer porch edge, setting that on posts or columns for support.

Shed roofs can also come in higher than the top plate, though the job is a bit more difficult. The shingles and the roof sheathing must be removed for the needed stretch up the house roof. The new rafters are then nailed to the house rafters and toenailed to the top plate, and new sheathing is placed. The outer end is done in the same manner.

Areas where new roofing goes on should be flashed with aluminum bent to make a small valley. The top edge of the flashing is inserted under the original shingles left on the roof,

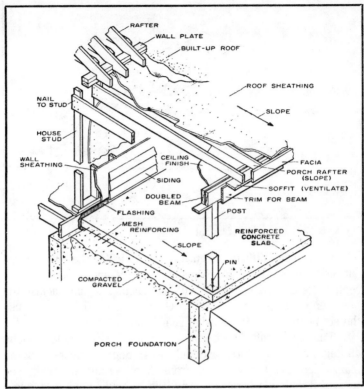

Fig. 12-5. Attaching a shed roof porch to the house.

and the bottom edge is placed over the shingles or built-up roofing on the new roof. This lower line is then sealed with hot asphalt, spread thin enough that it doesn't impede water flow.

Gable Roof

Gable roofs for porches are harder to erect, but far from impossible for the amateur who is willing to take a bit of care. The hardest part is getting the correct angles on the valley rafters and on the new ridge board. The porch, once framed, is measured for span and run, as you would with any standard building, to determine the rafter lengths needed. The ridge board is cut to fit the roof slope at the old roof. The rafters for the front end of the porch are measured and cut as described in Chapter 10. Figure the number of rafters needed to get to the old roof. Cut this number from the patterns formed by the first pair. Erect the ridge board, the first set of rafters, and the remaining full cut sets, making sure all are plumb (Fig. 12-6).

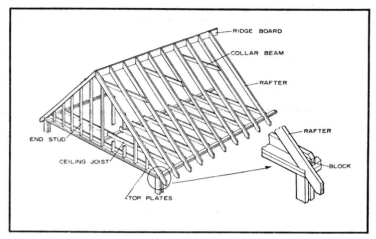

Fig. 12-6. Gable porch roof layout.

If you have difficulty figuring the angle for the ridge board cut, use a bevel T square to get the correct angle. Set it on the roof, place the ridge board, with its uncut end, and adjust the blade until the angle is correct. Once back on the ground, use the set angle to mark the ridge board and make your cut.

The valley rafters are cut to the appropriate angle (usually 45 degrees on each side) and half the 45 degree thickness of the main ridge board is removed from them. From this point, you will need valley jack rafters. Any jack rafter or stud is simply a shorter than normal piece. Again, the sliding bevel T square is the best way to make sure your angles are correct before cutting. Each rafter will be a different length than the one next to it. There will be two rafters cut to each pattern.

Plates must be laid along the old rafter line. It is always a good idea to lay out your gable roofed porch addition so that the plates will fall directly over existing rafters.

There are moderately complex methods for figuring the lengths and cuts for valley jack rafters. It will be much simpler to erect the full length rafters and ridge board. Then make the cuts as marked individually on 2 by 4s carried into place and checked. There is also less chance of making mistakes in measuring and cutting.

Once all the rafters are in place, the sheathing is installed, cut to the correct angles in the valleys. Before roofing can be applied, you will have to flash the valleys, using either metal flashing or shingles cut to fit. When two roof lines intersect, there is a valley formed. Some type of flashing is needed to prevent leaks. If the slope is at least 7 in 12, you can use flashing a foot wide. For slopes from 4 in 12 to 7 in 12, you must go to 18-inch wide flashing. If the slope of the roof is less than 4 in 12, the flashing must be 2 feet wide. The minimum open width of a valley is 4 inches at the top. It should widen at a rate of about ⅛ inch per foot as it falls down the roof. Placing a chalk line on the flashing before applying the shingles is an easy way to maintain this distance (Fig. 12-7).

GUTTERS AND DOWNSPOUTS

On most porches, gutters and downspouts will prevent too much moisture around the porch and, possibly, in the basement. The most common and about the easiest to install are the aluminum types. Some companies are now developing systems especially to make the job a bit easier for do-it-yourselfers with minimal experience. Essentially, the ease consists mainly in having the materials in easy to handle 10 or 12-foot sections (Fig. 12-8).

Most gutters are hung from the edge of the roof or connected to the fascia. They may be hung on straps run onto the roof or spiked directly to the eave ends of the rafters.

In determing gutter size, allow a square inch of downspout for every 100 square feet of roof surface to be drained. If the downspouts are spaced up to 40 feet apart, the gutter should have the same area as the downspout. It is unlikely on a porch that you will be draining much more than a few hundred square feet; nor is it likely you'll space the downspouts more than 40 feet apart.

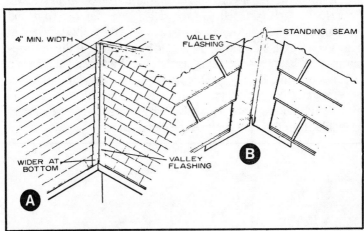

Fig. 12-7. Valley flashing details.

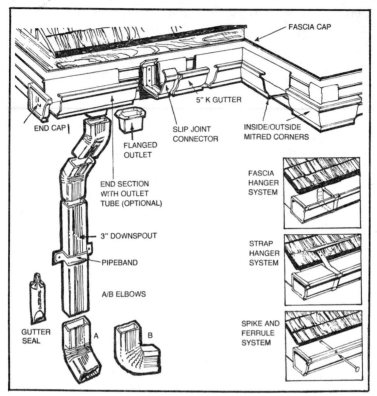

Fig. 12-8. Details of gutters and downspouts.

Aluminum hangers are spaced no more than 30 inches apart. Downspouts are fastened to the wall at least at the tops and bottoms, using straps which are then nailed to the wall. If you are in a high wind area, a central strap is also a good idea. If the downspout is extra long (over 10 feet), use straps at intervals of no more than 6 feet. An elbow at the bottom of the downspout will direct water away from the structure or onto a splash block. Joints on gutters and downspouts are sealed with a special mastic usually available from the manufacturer or distributor.

Chapter 13
Tilt-Up
Timber Buildings

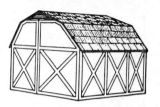

Span buildings of good size may not be of interest to everyone, but they can be useful to stock farmers. The building methods described here are not for the suburbanite, then, but for the farmer who wishes to erect span buildings in sizes large enough for use as hay storage areas, bull barns or cattle feeding barns. The technique is known as *rigid frame construction*. A rigid frame building is defined as one in which the connections between the roof and side walls are rigid. These angles cannot be changed without overstressing the structural timbers and their connectors (Fig. 13-1).

RIGID FRAME BUILDING

Figure 13-2 shows the basics of design for a rigid frame building adaptable to spans of as little as 26 feet and as great as 48 feet. Construction using rough lumber is reasonably cheap. As Table 13-1 shows, the increase in strength of the timbers over planed wood of the same nominal size is quite large. The design in Fig. 13-2 requires rough lumber in 1 and 2-inch thicknesses, with no single member required to be over 16 feet long. A great many figures are expressed as W, the width of the center section. Follow these dimensions closely, as they are critical to the strength of the building. Make sure columns are vertical and the crown fit is correct.

Frame Layout

Construction starts with the frame layout after the building size is determined. As Fig. 13-3 shows, a jig layout is used. It is best laid out on a wood floor, but fairly level ground will also serve if the wide planks are firmly staked down. Blocks are nailed to the wide planks and located so they control the alignment of the top and outside of the frame.

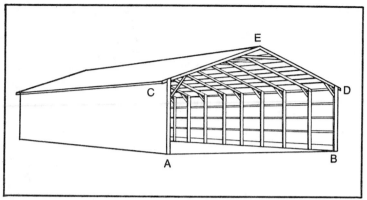

Fig. 13-1. A rigid frame building.

During layout, make only essential cuts which must be precise. It is usually easier for you to trim any protruding timbers after the layout.

All timbers must be securely held against the guide blocks until at least one nail is driven into each joint. Then you can move on and drive the needed number of the correct size nails in each joint. Drive half the nails from one side of the joint and half from the other.

For tilt-up construction, you must locate the 9/16-inch hole precisely in the same spot on each section laid out. This serves to hold the pivot after the wall section is assembled. Getting this out of line will make tilt-up nearly impossible.

Table 13-1. Percent Increase in Strength of Rough Lumber Over Dressed Lumber.

SIZE	BENDING STRENGTH	TENSIL AND COMPRESSIVE STRENGTH
2×4	50	36
2×6	40	31
2×8	40	31
2×10	36	30
2×12	34	28
1×4	36	42
1×6	46	37
1×8	46	37
1×10	42	35
1×12	40	34

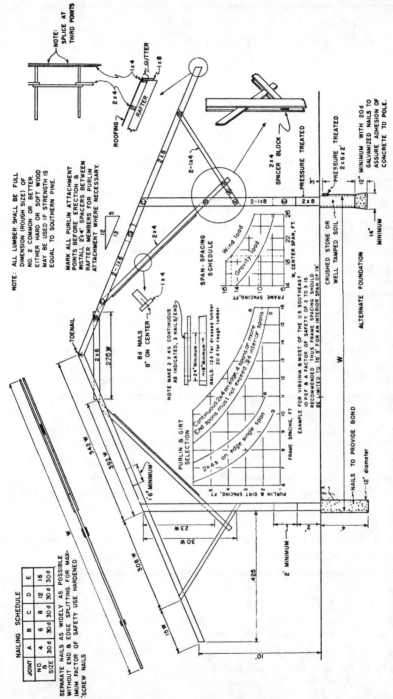

Fig. 13-2. The general design of an open frame building from rough lumber, to be 26 to 48 feet in span.

293

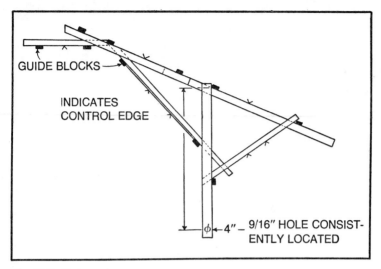

GUIDE BLOCKS

INDICATES
CONTROL EDGE

4" — 9/16" HOLE CONSIST-
ENTLY LOCATED

Fig. 13-3. Jig layout.

Foundation

As always, the building site is located or graded so that drainage will be away from the building. Use a post system for the foundation. Get the holes well into the ground, using concrete or other methods to make sure they are securely anchored at the correct depth. See Chapter 8 for pole depths needed, and make sure to add the height of the timbers to be tilted in place. Foundation timbers are located to match the timbers laid out in the jig, and 9/16-inch holes are again drilled at the exact same elevation in all of these timbers. Usually the holes are drilled about 3 feet above the ground level and 4 inches in from the edge of the foundation timber (Fig. 13-4).

Now the assembled half frames are set near the foundation timbers. The ends are lifted and ½-inch bolts are inserted through the holes in the half frame and the foundation timbers. These are not drawn up tight. The tops of the rafters are to be resting on the ground near the opposite side of the building.

A tractor or other vehicle is attached to the haunch of one half frame by means of a strong rope. The half frame is then pulled into place. It is plumbed and left tied to the hoisting vehicle. An opposing unit is bolted in place, roped and pulled up to be attached to the first half frame. Collar beams are placed and nailed securely before the hoisting vehicles are released to go on to the next pair of half frames.

Columns

Once all the sections are in place, with collar beams nailed and any needed girts on rafters and side walls added, the end columns should be carefully plumbed. Then intermediate frames are carefully aligned with the end frames and permanently nailed at the column/foundation joints. If wall girts are not used in open building construction, you'll have to add cross bracing between all the columns to aid rigidity.

Buildings erected according to these specifications will always have a two or three time safety factor over the design load. They can be strengthened further by the traditional methods of reducing column spacing, using lumber of better quality than number 2 Southern pine. Ring shank nails will also add strength and are recommended if green lumber is used.

CLOSED BUILDINGS

Closed buildings can also be constructed in much the same manner, with the possibility of getting an unobstructed span of as much as 44 feet using lumber no longer than 16 feet. Frame layout and jig construction is the same for closed buildings as it is for open buildings. Figure 13-5 shows some of the critical values needed to make the job a good one. Figure 13-6 shows the needs for a building of rough lumber, in the same size, while Fig. 13-7 illustrates the requirements for a closed frame build-

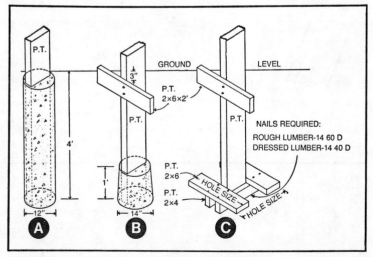

Fig. 13-4. Foundation choices and ground level line.

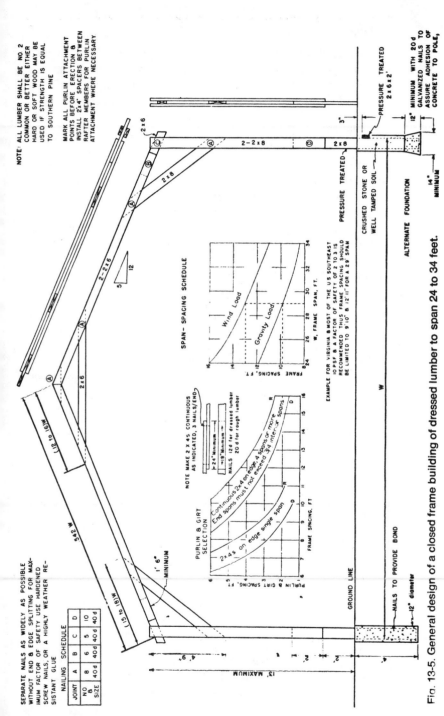

Fig. 13-5. General design of a closed frame building of dressed lumber to span 24 to 34 feet.

296

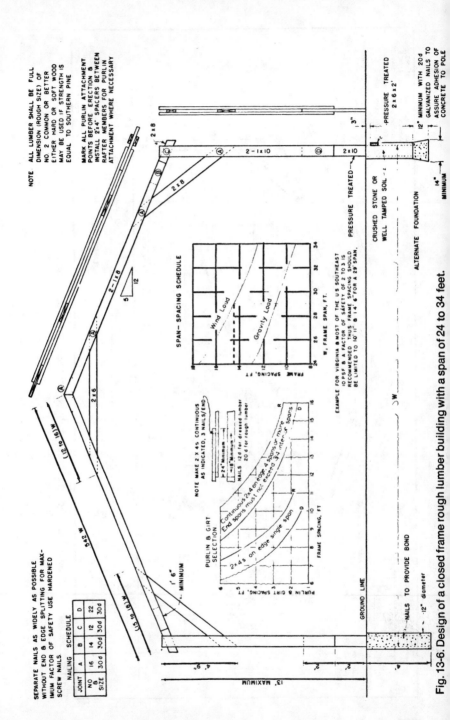

Fig. 13-6. Design of a closed frame rough lumber building with a span of 24 to 34 feet.

297

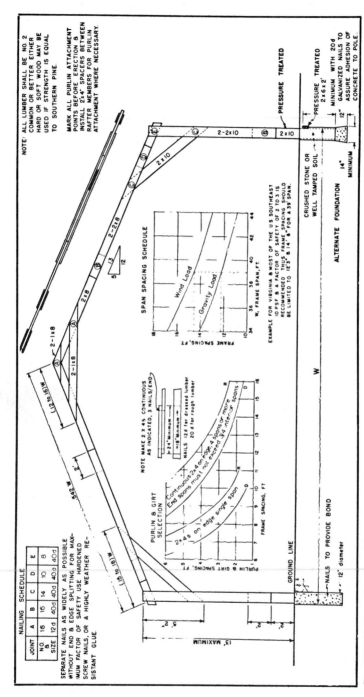

Fig. 13-7. Design of a closed frame building of dressed lumber for a span of 34 to 44 feet.

298

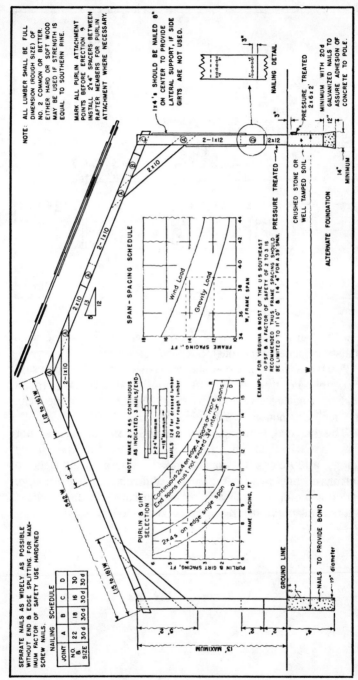

Fig. 13-8. Design of a closed frame building of rough lumber for a span of 34 to 44 feet.

299

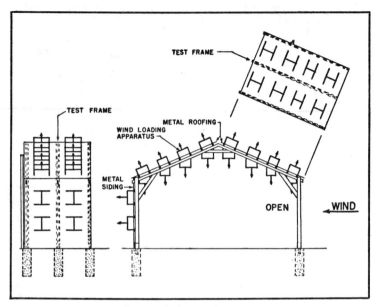

Fig. 13-9. Application of loads during testing.

ing with a span to 44 feet, of dressed lumber. The same span can also be gained with rough lumber (Fig. 13-8).

All these configurations shown have been built and tested at Virginia Polytechnic Institute in Blacksburg, VA, to 20 pounds per square foot roof loadings. The structures use plain shank nails and number 2 Southern pine (Fig. 13-9).

The original buildings were erected in 1962. Cost per square foot was then figured, for 5,000 square foot open frame buldings, at 39 cents. Obviously, the cost has probably risen to 10 or more times that figure, but it is difficult to imagine a large building going up for a lower cost if you have the time and the tractors available to use these erection methods.

Chapter 14
Greenhouses

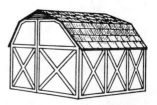

One of the more popular outdoor additions these days is the *greenhouse*. Some people simply want more room to grow houseplants, while others are having more and more trouble adapting to the plastic vegetable varieties we've come to know and hate over the past decade or two. There's nothing quite like having garden fresh tomatoes in mid-January. You name it, and you can grow it in a properly constructed greenhouse.

There are two ways to go about building your own greenhouse. The easiest is usually to buy a company kit model and then put it up. Many of these kits are complete from door to heater and will even include benches. All you need supply are pots, soil, seeds, water and fertilizer. While this method is usually at least 50 percent cheaper than having someone else build you a greenhouse from scratch, basic construction methods can easily be used for additions. A reasonably simple job results in a good, sturdy and productive home-built greenhouse.

The availability of translucent fiber glass in rolls adds to the ease of construction. The material is easily cut with a saw or tin snips and comes in rolls 3 or 4 feet wide. Fiber glass will block infrared rays, but in translucent form will let enough ultraviolet rays pass to allow good plant growth. Most of the designs here are simple enough to be erected in one or two weekends, with about the most difficult being the American Plywood Association (APA) design with which we start. It can be built in either an 8 by 8-foot size or in an 8 by 12-foot size.

APA GREENHOUSE

Basically, this green house requires seven panels of plywood and a variety of other materials, depending on the size you select for building. You can use either 10 mil poly sheeting or fiber glass sheet to cover the structure, with corrugated fiber

Table 14-1. Materials List for the American Plywood Association Greenhouse.

Recommended plywood: 303® textured plywood siding with desired surface texture, APA grade-trademarked.

A-C or C-C Exterior plywood, APA grade-trademarked for shelves.

PLYWOOD

QUANTITY	DESCRIPTION
5 panels	⅜ in. × 4 ft. × 8 ft. Exterior plywood siding
2 panels	½ in. × 4 ft. × 8 ft. A-C or C-C Exterior plywood

OTHER MATERIALS

QUANTITY	DESCRIPTION
80 lin. ft.	10, ⅜ in. × 6 in. × 8 ft. basket weave-type fencing
228 lin. ft.	2 × 4 lumber for framing
24 lin. ft.	1 × 2 lumber stiffening for shelves
30 lin. ft.	1 × 4 lumber for door casing
68 lin. ft.	2 × 2 lumber for door framing
12 lin. ft.	1 × 2 lumber for door insert framing
15 lin. ft.	1 × 2 lumber for door stop framing
As required	Plastic for seasonal use: 6 to 10 mil polyethylene. One continuous piece 12 ft. wide × 17 ft. long. Two pieces 8 ft. × 8 ft. for doors and ends.
Optional	For more permanent installation use corrugated fiberglass panels applied horizontally on the roof and down to the sides—applied with corrugations running vertically on the ends and doors.
2 pieces	Approximate size: 36 × 42 in. sections of screening
6	Door hinges (3 per door) 3-½" butt hinges
4	Vent hinges (2 per vent) 1-½" butt hinges
10	2 in. galvanized carriage bolts
40	1 in. round-head machine screws (washers not necessary)
As required	1 in. construction nails
As required	Caulk, sealant or flashing material to fill any joints, etc.
As required or desired	Finishing materials—stain or paint, wood preservative (see Building Hints)

NOTE: In areas of heavy snow, do not allow buildup on the roof.

glass panels on the roof. The lighter weight panels should easily bend to fit, or you can ignore them and stay with the sheet. Table 14-1 provides a materials list. While Fig. 14-1 and 14-2 give floor plans for both sizes. Figures 14-3 through 14-20 show layout and assembly details for various features, including the roof vent and all benches and shelves. Figure 14-20 also provides a look at arch assembly and gives you a good idea of what final overall appearance will be.

This greenhouse is fairly elaborate and might be beyond your needs or desires. Other forms are readily designed and can be cheaper and more easily heated and cared for than free-standing greenhouses. An attached lean-to greenhouse is one of those designs, but you should carefully consider a couple of things when planning to build a lean-to greenhouse.

While it is generally easier to provide light, heat and electrical circuits to a lean-to, and there are only three walls, at most, to erect, there are severe space limits. It is difficult to

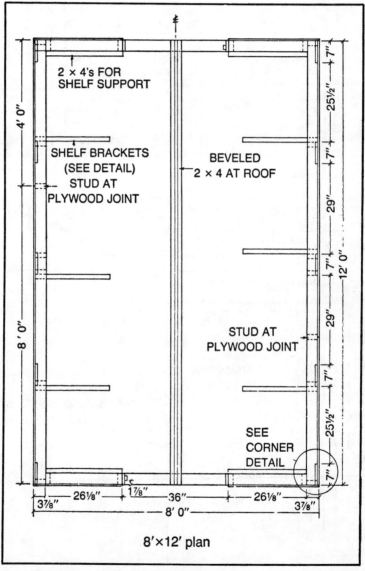

Fig. 14-1. Basic layout for the greenhouse.

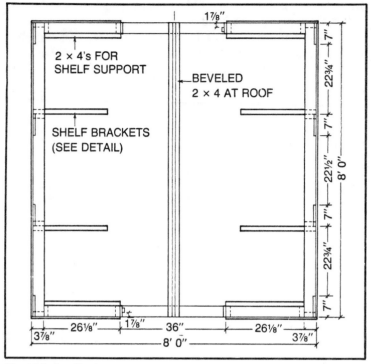

2 × 4's FOR
SHELF SUPPORT

BEVELED
2 × 4 AT ROOF

SHELF BRACKETS
(SEE DETAIL)

1⅞"

7"

22¾"

22½"

8' 0"

22¾"

7"

26⅛"   1⅞"   36"   26⅛"

3⅞"   8' 0"   3⅞"

Fig. 14-2. Smaller layout.

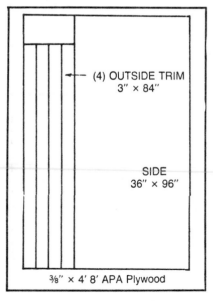

(4) OUTSIDE TRIM
3" × 84"

SIDE
36" × 96"

⅜" × 4' 8' APA Plywood

Fig. 14-3. Layout of the ⅜-inch
by 4 by 8-foot plywood sheet.

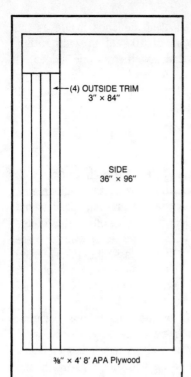

(4) OUTSIDE TRIM
3″ × 84″

SIDE
36″ × 96″

⅜″ × 4′ 8′ APA Plywood

Fig. 14-4. Layout of the next ⅜-inch sheet.

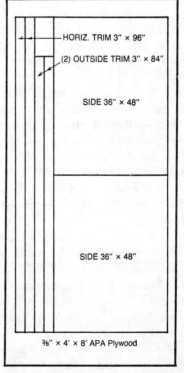

HORIZ. TRIM 3″ × 96″

(2) OUTSIDE TRIM 3″ × 84″

SIDE 36″ × 48″

SIDE 36″ × 48″

⅜″ × 4′ × 8′ APA Plywood

Fig. 14-5. A third required panel layout.

get more than a double row or planting benches inside as the lean-to is pretty much limited to 8 or 10 feet in width. Also, light is limited since it can only strike well from one side. It is almost essential that a lean-to greenhouse be built on the south or east side of a house or other structure. The lack of a fourth wall also may limit available ventilation, causing some problems with temperature control on hot days.

LEAN-TO GREENHOUSE

This design is 12 feet long and 8 feet deep. It can be built directly against a wall with a door if you wish to eliminate the need for getting, or building, a door for the greenhouse. I would recommend that you provide such a door even if the house opens into the greenhouse. Two vents on the long wall give adequate ventilation, and corrugated fiber glass sheets are used for the roof to speed matters up. This design, once the plan is adapted to your situation, should take no more than one weekend to put up. See Figs. 14-21 through 14-27.

Use pressure-treated 4 by 4s at the two outside corners, set well below ground frost level in concrete, if you don't want to lay a footing and 4-inch foundation wall. Attach a ledger board under the roof, bolting or lag screwing it directly to the house wall studs or top plate. Start 1 foot in from the end (approximately) and use one lag screw every 2 feet for good security. The lean-to outside height should be at least 6 feet. Even for taller people, height here isn't of overwhelming importance. There will be a planting bench to keep people back from the point where heads might strike rafters.

If 4 by 4s were used as corner posts, you can run a triple sole plate of pressure-treated 2 by 4s, with the bottom member set in a 4-foot bed of gravel. From this point, frame the walls on 2-foot centers, using a single or double top plate as your local codes specify.

Above the top plate, set a rafter in place. Use a bevel T square to get the angles needed. Cut the first—use 2 by 4s for rafters—and check the fit. Cut the rest from this pattern. Now cut at least two cripple studs for each end to fit from the top plate to the end rafters (end rafters should be doubled). Use Teco hangers to attach the rafters to the 2 by 6 ledger board, and then nail from the top of the angle cut into the top plate.

Cut two pieces to fit between the rafters to form a frame for the top vent. Then cut one more to frame the wall vent.

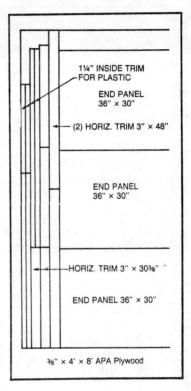

1¼" INSIDE TRIM
FOR PLASTIC

END PANEL
36" × 30"

(2) HORIZ. TRIM 3" × 48"

END PANEL
36" × 30"

HORIZ. TRIM 3" × 30⅜"

END PANEL 36" × 30"

⅜" × 4' × 8' APA Plywood

Fig. 14-6. Two more panels laid out.

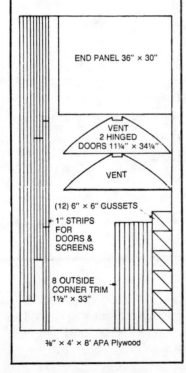

END PANEL 36" × 30"

VENT
2 HINGED
DOORS 11¼" × 34¼"

VENT

(12) 6" × 6" GUSSETS

1" STRIPS
FOR
DOORS &
SCREENS

8 OUTSIDE
CORNER TRIM
1½" × 33"

⅜" × 4' × 8' APA Plywood

Fig. 14-7. Many things will be cut from this piece of ⅜-inch by 4 by 8-foot plywood.

307

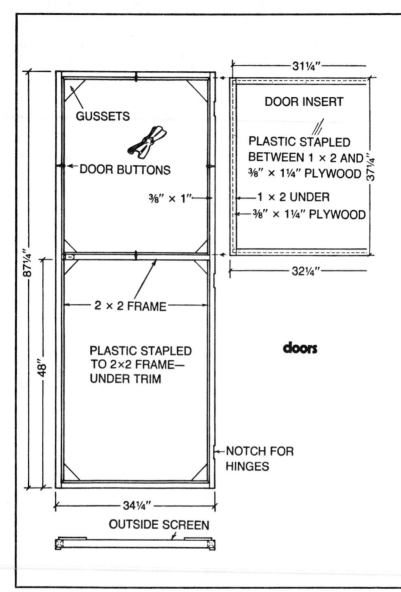

Fig. 14-8. Details of the greenhouse doors.

This vent uses the top bottom plate for a bottom part of the frame. Place the wall vent at the end of the long wall opposite the door, the top vent close to the top of the roof, and one rafter spacing in from the door for best air flow.

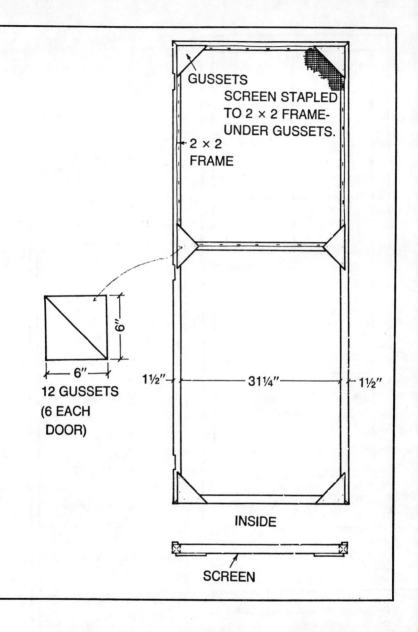

GUSSETS

SCREEN STAPLED TO 2 × 2 FRAME- UNDER GUSSETS.

2 × 2 FRAME

6″

6″

12 GUSSETS (6 EACH DOOR)

1½″ — 31¼″ — 1½″

INSIDE

SCREEN

Nail the corrugated roofing in place, using galvanized screw nails. A grommet or washer keeps the job waterproof as it seals the nail hole. You will probably need to place 2 by 4 fillers between the rafters along the top plate. A piece of corrugated

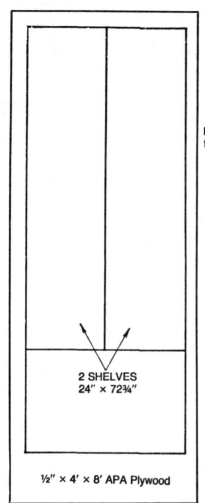

Fig. 14-9. The first shelves are cut from this ½-inch plywood sheet.

2 SHELVES
24" × 72¾"

½" × 4' × 8' APA Plywood

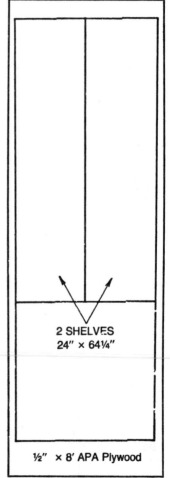

2 SHELVES
24" × 64¼"

½" × 8' APA Plywood

Fig. 14-10. Remaining shelves are cut from this sheet.

molding—available from the place selling you the fiber glass panels—on top of this will provide a good air seal. Leave about a 2-inch overhang, front and sides, for the roof.

The siding should be of fiber glass sheets or of the continuous sheeting already mentioned. This is nailed to the studs and is then covered with redwood, cedar lath, or pressure-treated lath.

The vents are simply made of 1 by 2s, nailed to be 2 feet high and 22½ inches long. For the roof vent, use the same corrugated material as you did for the roof and the same type of molding between the rafters to get an air seal. For the wall vent, frame to the same size. Use the flat sheeting, again covering the nails with lath.

The door is a bit harder, but by using butt joints it is easily framed. Brace with Teco metal braces, making sure the braces

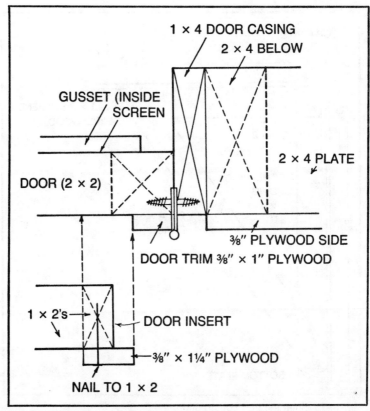

Fig. 14-11. Details of hinges.

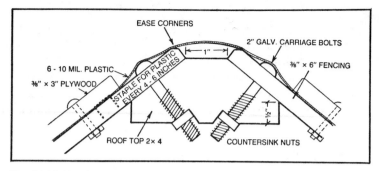

Fig. 14-12. Roof peaks details.

are galvanized. Make the door to fit the opening you have framed for it. Cover the door with the fiber glass sheet. Add hinges, hang carefully, move in some planting benches and start work.

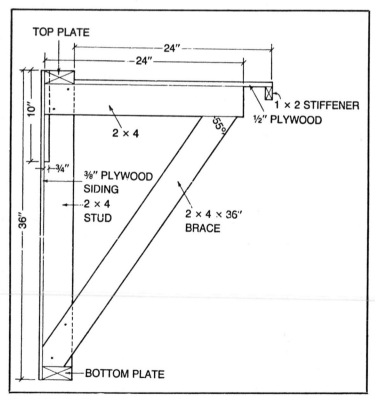

Fig. 14-13. Shelf support construction.

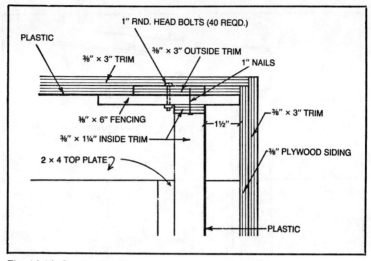

Fig. 14-14. Corner details.

GABLE ROOF GREENHOUSE

A freestanding greenhouse with a gable roof can be placed on any portion of your lot that gets enough sun to grow the plants you want. Use the basic small building construction methods already covered, taking into account that this buildings is lightweight (Figs. 14-28 and 14-29). A good footing and foundation is a help if you wish to keep the greenhouse standing for any length of time. But the foundation doesn't need to be more than 6 inches wide if you can find block in that dimension locally. Lay sill plates and anchor as usual. Continue on up, using

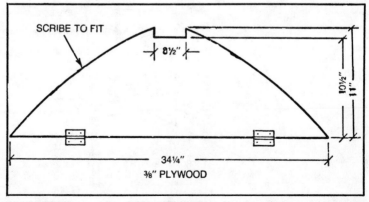

Fig. 14-15. Construction of the ventilator.

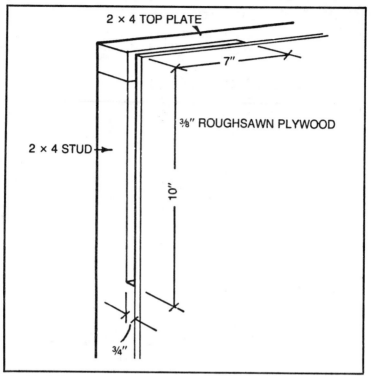

Fig. 14-16. Notch details at corners.

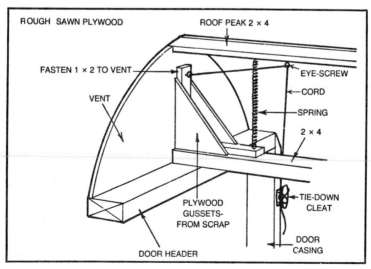

Fig. 14-17. Assembly details for the vent.

2-foot on-center stud and rafter spacing to allow for greatest light penetration.

Use fillers between the rafters and corrugated sheets, with corrugated fillers, for roofing. Again, sheeting applied to the side walls is the easiest way to go. Weather-resistant wood can cover the nailheads on the studs.

This type of greenhouse is a bit harder to get water, electricity and heat to than a lean-to type. But it is larger and has much more adaptability regarding the site.

For any greenhouse, it is a good idea to line the interior walls with at least 6 mil poly sheet. This does not significantly cut down on light transmission, but does provide an air barrier that will help reduce heat loss. Have a helper hold the roll while at one end you nail, through lath, the poly sheet in place. It should then be stretched as tightly as possible across the studs and nailed, always through lath, to each stud. You can do the

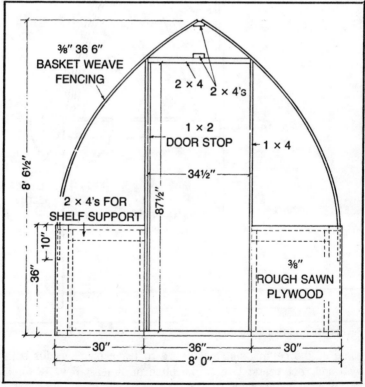

Fig. 14-18. Check dimensions.

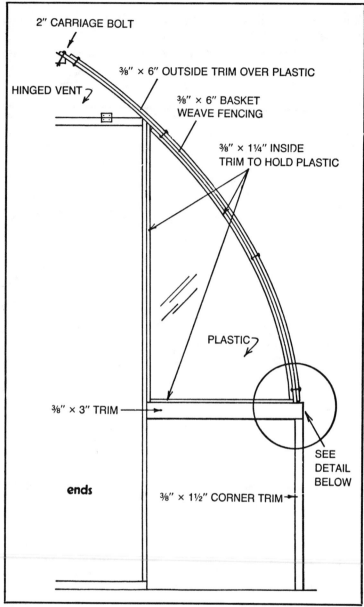

Fig. 14-19. End fastening details.

same at the ceiling, but make sure to leave openings for both wall and roof vents. Use a roof pitch of at least 6 in 12 for a greenhouse with a gable roof.

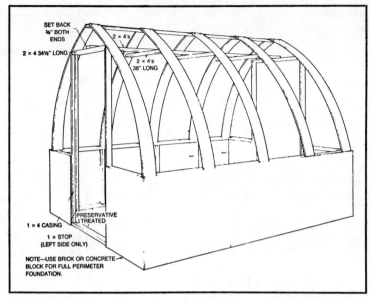

Fig. 14-20. Completing the arch assembly shows you the finished form of your APA greenhouse.

HEATING

Almost any heat source can be used in a greenhouse, though a wood stove in a plastic and frame greenhouse must be set up properly to be safe. Soil heating cables under the plant beds are often a good idea.

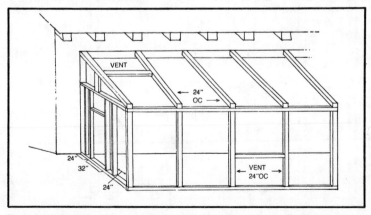

Fig. 14-21. An overall look at the lean-to-greenhouse framing. Length can be determined by your needs.

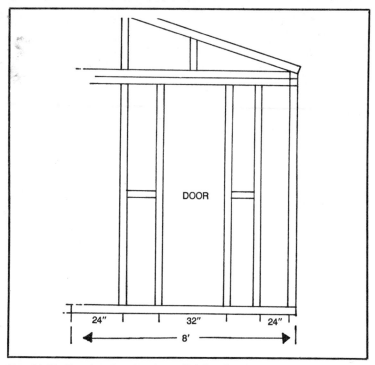

Fig. 14-22. Framing for the door. Height should be at least 74 inches whenever possible.

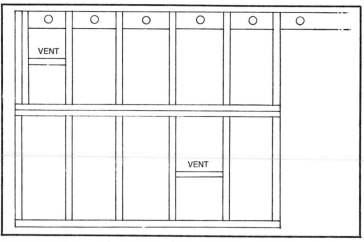

Fig. 14-23. Framing is on 24-inch centers. The ledger is bolted to the house studs or upper plate every 24 inches on center also, using 3-inch lag screws. The joists are hung with Teco joist hangers.

318

Figure your heating needs on the maximum desired indoor temperature at night—say 60 degrees Fahrenheit. Look up the lowest expected readings for your area of the country in midwinter. Let's say that figure is 5 degrees Fahrenheit. That is a temperature difference of 55 degrees Fahrenheit. Now you need to know the square footage of exposed glass area in your greenhouse. We'll assume a smaller greenhouse with 1,500 feet of exposed glass area. That is multiplied by the temperature difference, resulting in a total of 82,500. A single layer of glass or plastic will need a further multiplier of 1.2, while a double layer requires one of .8. The worth of the second layer is immediately apparent when you see that the British thermal unit (Btu) need for the single layer greenhouse is 99,000, while the double layer requires only two-thirds that amount, or 66,000 Btus.

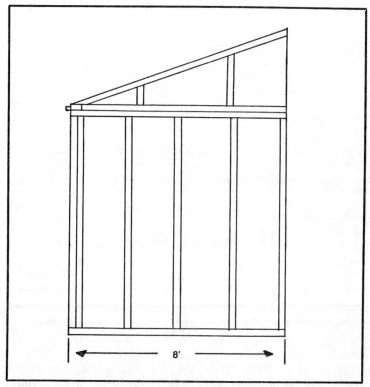

Fig. 14-24. End wall framing. The top plate should be doubled. Overall width on lean-to greenhouses should not be much more than 8 feet, and the long side should face to the south or southeast.

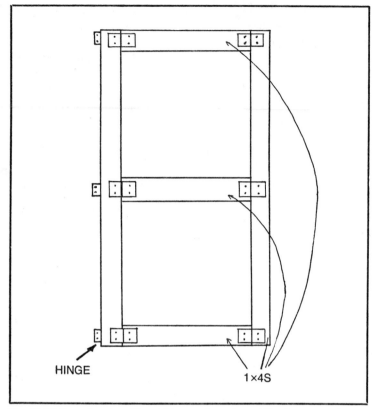

HINGE

1×4S

Fig. 14-25. Door framing. The width must be at least 30 inches, and it is best ot have a door height of 74 inches or more. Use Teco connectors to brace the door.

VENTILATION

Even on cool or cold days, a greenhouse may heat up too much if the sun shines brightly. For this reason, all greenhouses need *ventilators*. In all the designs here, the ventilators are manually operated. Automatic ventilators are quite expensive. Automatic fans can also be used, and are practical if the greenhouse you build is fairly large. For the fans to be effective, they must be capable of a complete air change every minute. Thus, the capacity of the fan in cubic feet per minute, at ⅛-inch static pressure, is a factor you need to know. For most greenhouses, you can determine the volume by simply multiplying the floor area by seven. A fan at each end, and as high in the end walls as possible, is most effective.

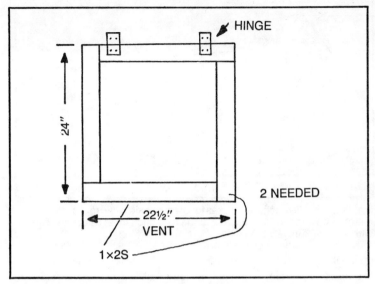

Fig. 14-26. Vents are framed as shown with 1 by 2s.

Occasionally, the greenhouse will need protection from too much sun. You'll find a variety of shades available for both the interior and exterior.

To increase yields, you may have to find some method of enriching the carbon dioxide levels when the vents are closed

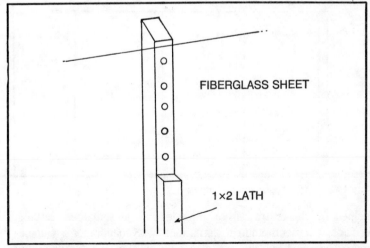

Fig. 14-27. Use lath to protect nail heads and to help prevent possible tearing of the fiber glass sheet. Use only screw nails with grommets to fasten the sheet.

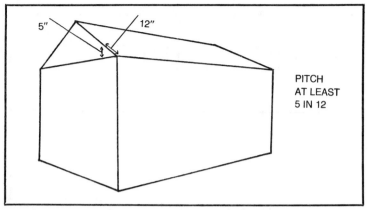

Fig. 14-28. Gable roof greenhouses follow the basic gable roof construction methods covered in Chapter 10, but the roof pitch must be at least 5 in 12, with at least 6 in 12 preferred.

on cool days. You can get inexpensive kits to monitor the carbon dioxide levels. Then either buy bottled carbon dioxide or get special equipment using infrared sensors and electric automatic release to take care of the job for you. Plants especially

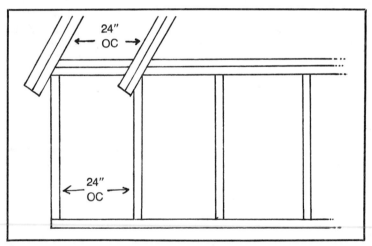

Fig. 14-29. 24-inch on-center framing is used.

helped by this extra carbon dioxide include tomatoes, lettuce, orchids, chrysanthemums, carnations and snapdragons. Supplemental electric lighting will also help on short winter days, but it should be of high intensity.

Chapter 15
Concrete Projects

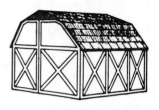

We've already covered some uses of concrete, and the methods of curing, selecting aggregates, mixing and other such chores. There are a fair number of places around a home or farm where concrete can come in handy, though, which have yet to be examined. For a great deal of the work described in this chapter, some form of reinforcement of the concrete will be needed for long life. Either round bars or steel mesh can be used. The mesh is usually more popular for concrete slabs such as drives and walks, while the bars are more likely to be used in walls and stairs. In any case, steel should be free of dirt, oil, scaling, rust and grease. If any of these remain present in a major degree, bonding will not be good and reinforcement will be lessened.

WALKS

The width of a concrete walk will depend on your needs. Generally, even the narrowest should be 2 feet wide, with few found wider than 6 feet. Walks do not, generally, need reinforcing material. In most areas a 4-inch slab thickness will be sufficient if laid over a good base. For heavy duty use, especially when there is a chance of large vehicles crossing the walk, a 6-inch slab using wire mesh reinforcement is a good idea.

The base for the walk should be of gravel, tamped well, in a layer at least 2 inches thick. Forms are laid on this base, using boards exactly the width that you wish the concrete to be deep. Provide slip boards in the forms every 4 to 6 feet. These can be pulled out to provide expansion joints since no footings are used. The individual slabs must be allowed some room to move to prevent cracking (Fig. 15-1).

If you wish to prevent water from standing on the walk, build one side of the walk slightly higher than the other if the

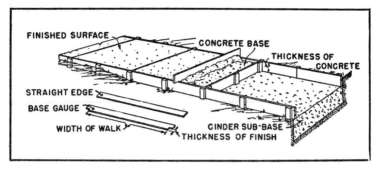

Fig. 15-1. Sidewalk forms for concrete.

ground is otherwise level, or nearly so. Crowning is also possible, but is a bit more work and provides more room for error (Fig. 15-2).

Leave the walk after the first wood float finish if it is level. On upgrades it is a good idea to take a stiff broom and make crosswise score marks to aid in traction on wet days (Fig. 15-3).

STEPS

Poured concrete steps are considered a fairly complex job by many people. The need for estimating the amount of mix required scares many folks. The easiest way to determine the cubic yards of concrete needed for concrete stairs is to consider

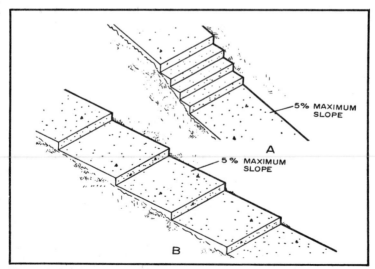

Fig. 15-2. Slope to one side no more than 5 degrees for drainage.

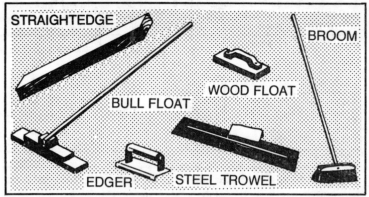

Fig. 15-3. Tools for working concrete.

each riser/tread unit as a separate slab. Figure the volume for each slab and add together to get the total. If the steps are earth-supported, forms such as those shown in Fig. 15-4 are used. For such steps, assuming the step width doesn't go past 4 feet and there are no more than four steps to be poured, slab depth need only be 4 inches.

For longer flights and wider steps, use a slab depth of at least 6 inches. These steps will be self-supporting. When no solid earth support is available, it becomes necessary to construct much heavier forms and to reinforce the concrete with steel bars. Table 15-1 provides the needed data on reinforcing bars for concrete slab lengths and thicknesses, as well as the spacing for the bars.

Longitudinal and transverse rods must be wired together. The longitudinal rods will run from the top to the bottom of the

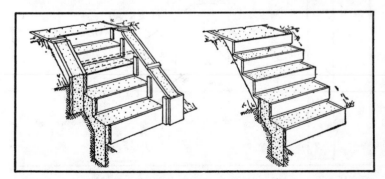

Fig. 15-4. Earth-supported steps can use these types of forms, shown here partly cut away to depict details.

Table 15-1. Data on Reinforcing Bars for Concrete Slab Lengths and Thicknesses.

SLAB DIMENSIONS		LONGITUDINAL RODS		TRANSVERSE RODS	
LENGTH (FEET)	THICKNESS (INCHES)	DIAMETER	SPACING	DIAMETER	SPACING
		Inches	Inches	Inches	Inches
2 to 3 -----------------	4	¼	10	¼	12 to 18
3 to 4 -----------------	4	¼	5½	¼	12 to 18
4 to 5 -----------------	5	¼	4½	¼	18 to 24
5 to 6 -----------------	5	⅜	7	¼	18 to 24
6 to 7 -----------------	6	⅜	6	¼	18 to 24
7 to 8 -----------------	6	⅜	4	¼	18 to 24
8 to 9 -----------------	7	½	7	¼	18 to 24

slab, spaced as Table 15-1 shows. The transverse rods must run the entire width of the slab or step being reinforced. Such self supporting steps require very solid support at the top end, usually a concrete or masonry wall. The bottom end will require a footing of good size, running to below frost depth.

For safest outdoor use, it is best to keep the risers no higher than 6 or 8 inches, with 6 preferred. Let the treads move out to 10 or 11 inches deep. Allow a slope of about 1/16 inch to the front of the tread to provide good water drainage from the steps. Figure 15-5 shows the type of forms needed, as well as the method of bending the longitudinal reinforcing rods for best strength.

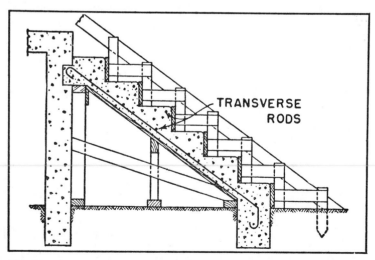

Fig. 15-5. Self-supporting steps require these forms of heavier construction.

Table 15-2. Spacing of Reinforcement Rods for Concrete Porches.

WIDTH OF PORCH (FEET)	SLAB THICK-NESS	REINFORCING RODS	
		DIAM-ETER	SPACING
	Inches	Inches	INCHES
4 ----------	5	⅜	7½
6 ----------	5 -	⅜	6
8 ----------	5½	½	9½
10---------	6	½	8

The transverse reinforcing rods are ⅜-inch round rods spaced 8 inches on center.

PORCHES

Concrete above-ground porches are a bit more complicated to build. The forms must be of heavy lumber and exceptionally well braced to prevent sagging. First, two 8-inch concrete block walls must be laid up, going down below frost depth, with footings at least a foot wide. Forms are made of 2-inch rough lumber for the bottom, with 2-inch finished lumber for the sides. Support must be provided about every 2 feet under the floor forms, usually in the shape of posts. More 2-inch lumber is jammed in place by the posts at 90 degree angles to the actual form. A slope of ¼ inch to the foot is maintained to provide drainage. Reinforcing rods, ⅜-inch in diameter, are spaced according to Table 15-2. See Fig. 15-6.

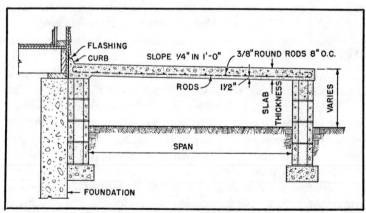

Fig. 15-6. Concrete porch construction details.

In all concrete construction, proper curing methods must be followed. A minimum time is seven days for stairs and porches. Walks can get by with a three day cure.

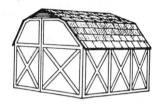

Chapter 16
Wiring

Since the subject of the book is outdoor buildings, I'm not going to give complete directions on wiring a home. What I'll look at is a type of wiring used often on farms, which is of some use in other areas. All wiring details for wire size and interior wiring will follow the National Electrical Code. It is always imperative to check local codes, as they are often more strict than the National Electrical Code (NEC). Many localities do not allow you to do your own wiring, while others require inspection by a licensed electrician or building inspector.

If you are allowed to do some wiring, check every circuit involved in the job to be sure it is cut off. Never work around or with live circuits.

Most outbuildings are unwired or wired only for a single outlet or light. This reduces utility a great deal and means no chance at using the structure as a workshop, studio or second office.

SERVICE ENTRANCE

A yard pole distribution setup is best if there is much distance between buildings and more than a couple of buildings to wire. The yard pole will usually provide a separate three-wire feeder service to each building to be electrified. The pole will be located centrally so that low voltage isn't a problem in buildings located a long way from the service entrance. Installation of the yard pole is up to your local power company (Fig. 16-1).

No service entrance should be less than 60 amperes, even if you only intend to use a circular saw or electric drill inside the building. One hundred amperes is better. If nothing else, going to this size service allows you to later wire another small building from the first, using an auxiliary panel. For workshop use, where major power tools are to be operated, I would recommend that at least 150-ampere service be installed.

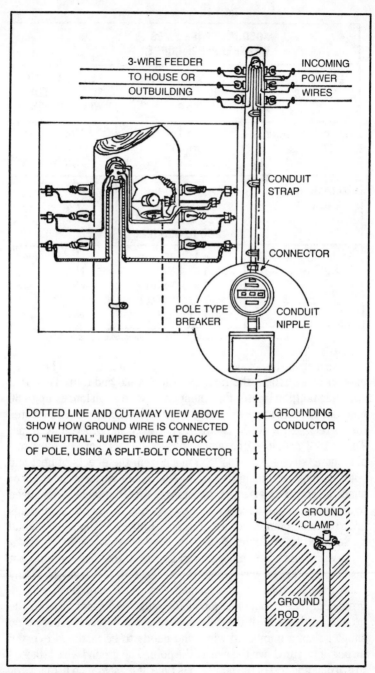

3-WIRE FEEDER
TO HOUSE OR
OUTBUILDING

INCOMING
POWER
WIRES

CONDUIT
STRAP

CONNECTOR

POLE TYPE
BREAKER

CONDUIT
NIPPLE

DOTTED LINE AND CUTAWAY VIEW ABOVE
SHOW HOW GROUND WIRE IS CONNECTED
TO "NEUTRAL" JUMPER WIRE AT BACK
OF POLE, USING A SPLIT-BOLT CONNECTOR

GROUNDING
CONDUCTOR

GROUND
CLAMP

GROUND
ROD

Fig. 16-1. Details of yard pole wiring.

329

Table 16-1. Adequate Wire.

ADEQUATE WIRE SIZES . . WEATHERPROOF COPPER WIRE		
LOAD IN BUILDING AMPERES	**DISTANCE IN FEET FROM POLE TO BUILDING**	**RECOMMENDED SIZE OF FEEDER WIRE FOR JOB**
Up to 25 amperes, 120 volts	Up to 50 feet 50 to 80 feet 80 to 125 feet	No. 10 No. 8 No. 6
20 to 30 amperes, 240 volts	Up to 80 feet 80 to 125 feet 125 to 200 feet 200 to 350 feet	No. 10 No. 8 No. 6 No. 4
30 to 50 amperes, 240 volts	Up to 80 feet 80 to 125 feet 125 to 200 feet 200 to 300 feet 300 to 400 feet	No. 8 No. 6 No. 4 No. 2 No. 1

These sizes are recommended to reduce "voltage drop" to a minimum

No matter the service size, the wire leading to it must be heavy enough to withstand the ravages of ice, wind and rain. The NEC specifies nothing lighter than number 10 wire at distances up to 50 feet, and nothing under number 8 over 50 feet. Table 16-1 indicates the correct size wire for loadings in both 120 and 240-volt systems. Table 16-2 provides the ampacities of copper wire. Never use aluminum wire. It "creeps" too much from the effects of heat and cold, causing terminals to loosen. Aluminum wire corrodes if not anodized and can cause high resistance. The first table is best used to determine needed feeder wire sizes when two or more buildings are to operate off one service entrance. Entrance cable sizes should be selected from Table 16-2.

All feeder wires must be run so that they touch nothing but insulators. Feeder wires must run at least 10 feet away from any obstructions, such as roofs and trees. The yard pole will be grounded. The grounding conductor must be copper, run to the ground on the side of the pole opposite the meter. It must be no smaller than a number 6 wire, and needs to be fastened every 6 inches of its run down the side of the pole. The ground wire attaches to a grounding rod of the nonferrous type at least ½ inch in diameter

Table 16-2. Ampacities of Copper Wires.

WIRE SIZE	IN CONDUIT OR CABLE		IN FREE AIR		WEATHER PROOF WIRE
	TYPE RHW THW	TYPE TW, R	TYPE RHW THW	TYPE TW, R	
14	15	15	20	20	30
12	20	20	25	25	40
10	30	30	40	40	55
8	45	40	65	55	70
6	65	55	95	80	100
4	85	70	125	105	130
3	100	80	145	120	150
2	115	95	170	140	175
1	130	110	195	165	205
0	150	125	230	195	235
00	175	145	265	225	275
000	200	165	310	260	320

Types "RHW," "THW," "TW," or "R" are identified by markings on outer cover

and 8 feet long. This grounding rod is located 2 feet or more from the yard pole and driven in until the top is a foot below ground level.

TAPPING

It is possible when a yard pole is used, to tap two-wire service from a three-wire and run the two-wire service into a small building. Usually the tap is made on wires running from the yard pole, making sure that the wire used for the tap is heavy enough to both carry the

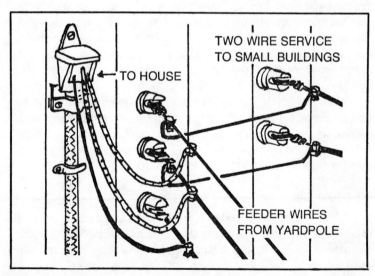

TWO WIRE SERVICE
TO SMALL BUILDINGS

← TO HOUSE

FEEDER WIRES
FROM YARDPOLE

Fig. 16-2. Three-wire service tapped to feed two-wire service to a small building.

331

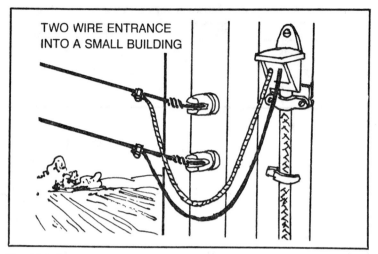

Fig. 16-3. Two-wire entrance to a small building, including drip loop and service head which are both required.

load and withstand the weather. Such a tap should not be used to any building requiring a service load of more than 3500 watts.

Use regular weatherproof wire of good quality; 8-gauge is best. Place the insulators on buildings so that the wires in the circuit are spaced at least a foot apart. Make sure the wires clear the ground by at least 18 feet over drives and 10 feet over walkways. Conduit or service entrance cable must be used on the vertical run. You need a drip loop on each wire before it enters the service head. Use a 30-ampere, 120-volt service panel, making sure fuses or circuit breakers match the capacity of the smallest wire (Figs. 16-2 and 16-3).

For other areas, underground wiring is sometimes a better bet than overhead wiring. Such wire should be of trench cable and needs conduit for protection if there is any chance of mechanical damage. All underground cable is best laid at least a foot below frost line to prevent heaving and condensation (Fig. 16-4).

WIRE AND CABLE

Most home wiring today is done with nonmetallic sheathed cable. There are two types in general use, NM and NMC. NM cable has paper wrapped inside. It has a plastic cable sheath and must be used only in dry locations. NMC has its individual wires encased in plastic before the cable sheathing is put on, so it can be used in wet or dry areas. As you might guess, NMC is more expensive than type

NM. Service entrance types have slightly different designations, but you can get what you need by asking your dealer.

Cable with two wires will have one white and one black wire. When the cable contains three wires, the third wire will be red. Ground wires are either bare, or insulated in green or with green stripes. Under no circumstances are fuses or circuit breakers ever placed in a ground wire circuit.

Conduit is tubing or pipe through which cables or wires can be passed. Rigid conduit will most often be of galvanized steel. EMT is thin wall metal conduit and is cheaper than rigid conduit, as well as being easier to work with.

Most of the boxes you'll need in outbuildings will be wall switch and receptacle boxes, light fixture boxes and junction boxes. Most are readily available in both plastic and metal and come in a wide variety of sizes. Check locally to see if plastic boxes are allowed. Remember, the NEC requires the use of a box any time you splice wires or connect wires to themselves or terminals. Boxes cannot be free swinging but must be correctly supported within the structure being wired.

All receptacles must be grounded, so that the receptacle will have three terminals. The green one is for the ground. Switches and receptacles today are easier than ever to hook up. Many are designed so that the stripped end of the wire need only be firmly pushed into the terminal to make a good connection.

STRIPPING WIRE

While wiring an outbuilding is not really difficult, some techniques provide for a neater and safer job. You may not be living in

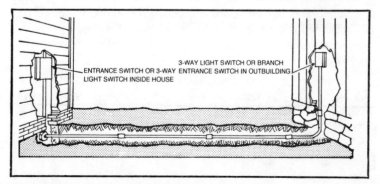

Fig. 16-4. Two-wire service run underground must be at least a foot below frost depth.

the structure, but there is little point in working to construct the building if it burns down the third time you flip a light on.

Stripping wire can be done with an electrician's wire stripping tool or a pocket knife. If the stripping tool is used, simply select the proper sized slot for the wire being stripped and do the job. A pocket knife requires more care, but can be extremely useful on larger wires and when you don't have a stripping tool for smaller wires. Don't use the knife to cut a circle around the insulation, nicking the wire. Instead, shave the wire away from you and toward the end, using care not to bite into the copper. Stranded wire needs even more care, as those tiny strands are very easily cut with a knife blade (Fig. 16-5).

Splicing these days is almost universally done using wire nuts. I still prefer soldered splices. But some local codes say you can't use them.

WIRING RECEPTACLES

Receptacles are easily wired these days. Because nonmetallic sheathing is more flexible than the old BX cable, the cable is run to the secured box. Check the receptacle. If it is the newest type, with push-in connections at the terminals, you need only strip back the insulation on the wire ⅜-inch and slip it in. Place the hot (black) wire on the brass terminal and the white or neutral wire into the silver terminal. The ground wire goes into the green terminal. Tighten the box clamp around the cable, screw the receptacle to the box, twist on the cover and you're done.

If your receptacle uses screw type terminals, strip the first ½ inch of the wires. Use a pair of needlenose pliers to make a partial loop in the wire. Screw out the terminal and slip the loop on. The open end of the loop goes in the direction the screw tightens (Fig. 16-6). Switch wiring is the same as receptacle wiring.

SAFETY

Never under any circumstances work on a circuit that is powered. Always use a circuit tester to check and make sure the power

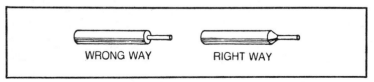

WRONG WAY   RIGHT WAY

Fig. 16-5. Right and wrong methods of stripping insulation from wiring.

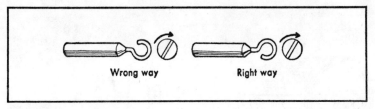

Fig. 16-6. Right and wrong ways of putting a loop on a screw terminal.

is off, even if you've pulled the fuse or cut off the circuit breaker. If you touch the wrong circuit breaker or fuse, the shock could be fatal.

Follow all codes explicitly. They are designed to provide you with safe electrical service.

Take your time. Electrical wiring is one craft where neatness does count. Neat wiring increases efficiency, and nick-free wires cut down on the chances of fires. If you are at all unsure of your work, even though local codes don't require inspection, pay a professional to look things over for and make suggestions or any needed changes.

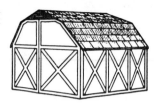

Chapter 17
Wood Kit Buildings

Outbuilding kits are exceptionally popular as anyone can see in a drive through a suburban area. Most are metal and offer the advantages of easy erection and general assembly, plus wide availability and low cost. There are some disadvantages to almost all these low cost buildings, though. One is fragility in high winds. An anchor kit of some kind is just about essential, and even these don't always work well. I have seen several small metal utility sheds scattered over a fairly wide area after a windstorm, and every one was a building the owner had thought pretty secure. If you don't put up a supplementary interior wood frame, the only shelving and other such conveniences that will fit are those sold by the manufacturer or distributor. Once anchored, floored and equipped, the low cost of these buildings often disappears. The first bad windstorm may make the building do the same.

ADVANTAGES OF WOOD FRAME BUILDING KITS

I have a strong preference for wood frame buildings. Anchoring is easier. Conveniences can be added at will, to your own design and desires. Expansion is also generally easier. Cost is higher, but one company has developed a method to keep prices down a bit, while still providing the advantages of wood construction. Jer Manufacturing has developed a kit that is shipped without the plywood and other costly to transport materials, which you then buy locally to complete the building. Rafters and trim pieces are supplied in precut form. Templates needed for front and back panels are also provided, as are nails, cap shingles and door hinges.

Four basic kits are available, with three gambrel roof barn designs and a salt box style carriage house design. All are 8 feet high and come in several sizes. For example, the budget barn comes in 8 by 8-foot and 8 by 12-foot sizes. The standard barn can be had in 8 by 8, or in lengths to 16 feet. The 10-foot wide barn also comes in

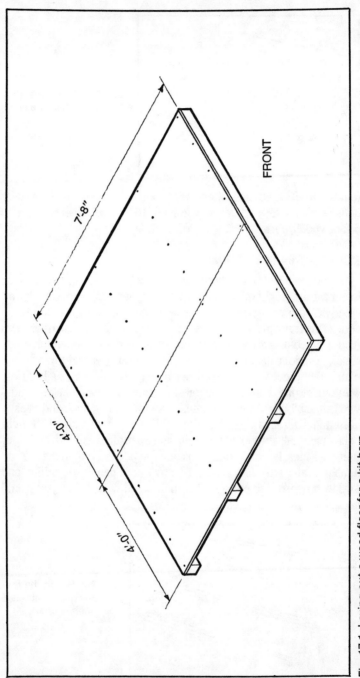

FRONT

7'-8"

4'-0"

4'-0"

Fig. 17-1. Laying out a wood floor for a kit barn.

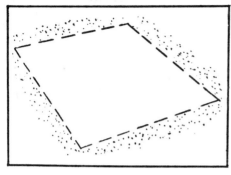

Fig. 17-2. Level the area where the barn is to stand.

lengths to 16 feet, and offers an 8-foot wide door. The carriage house style comes in different widths to 16 feet, all 8 feet deep, with 6-foot wide doors.

EIGHT-FOOT BARN MATERIALS

The standard 8-foot barn package includes the following: four lower rafters, eight middle rafters, four upper rafters, one back wall cross brace, one lower stud for the back wall, one upper stud, top door trim (two pieces), center horizontal and vertical door trim (three pieces), bottom door trim, outside door trim, door studs (two), one door header, a center door stop, upper and lower gable trim (eight pieces), outside corner trim (eight pieces), eight lower wall studs, one front wall upper stud, four side wall backer studs, two gauge blocks, 16 cap shingles, six hinges, two hook and eyes, a hasp and swivel, a bag of screws, 722 six penny nails, 108 eight penny nails, 52 sixteen penny nails and 35 roofing nails. In addition, there is a clearly drawn plan book and the needed templates. Your lumber dealer will then supply 10 4 by 8-foot panels for siding, and 11 2 by 4 by 92-inch purlins. For an optional wood floor, you would need four pressure-treated 4 by 4s 8 feet long, two sheets of

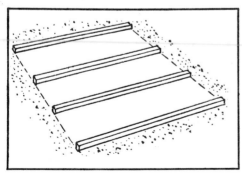

Fig. 17-3. Space the four runners equally, about 2 feet 7 inches apart.

Fig. 17-4. Cut flooring to the size indicated in Fig. 17-1.

exterior grade plywood (at least ⅝-inch thick) and ½ pound of 8 penny nails. For the shingled roof, four bundles will do, along with 4 pounds of 1-inch galvanized roofing nails.

Add to this a hammer, saw, level, tape and screwdriver (Phillips) and you're ready to go. Actually, one or two more pressure-treated 4 by 4s for use as piers, some 10 by ½-inch carriage bolts and enough ready-mix concrete to anchor them would be a good idea. You'll then need a drill and an electrician's bit (½-inch) to drill the piers to match the floor beams. To do the anchoring job properly, I would place a pier at the outside edge of each of the four floor cross members on each side of the building, while still allowing the four runners to rest on leveled ground. Such a procedure will prevent a windstorm from blowing the building away and will keep any frost heave at bay, assuming you go below frost line for your piers and anchor them with concrete.

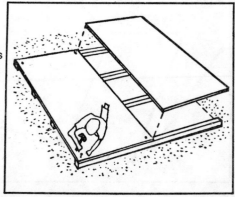

Fig. 17-5. Nail the two sheets of flooring to runners. Use 8 penny nails. Keep front and side edges flush.

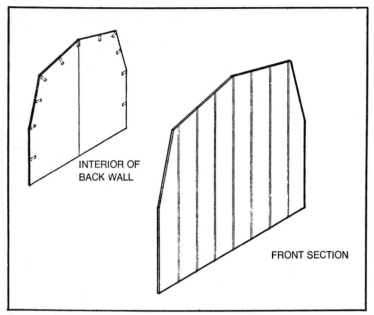

Fig. 17-6. View of the barn's front section.

BARN CONSTRUCTION TECHNIQUES

Once the materials and tools are gathered, lay out for the piers and dig the holes. Set the piers and pour the concrete. Give it a few days to set. Then place the runners against the piers and use the electrician's bit to drill the holes to accept the carriage bolts. Drive the carriage bolts through the holes, palce a washer on the nut end and tighten.

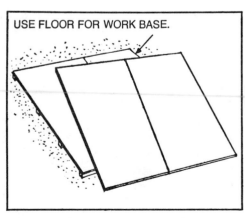

Fig. 17-7. Butt two full sheets of siding tightly together.

340

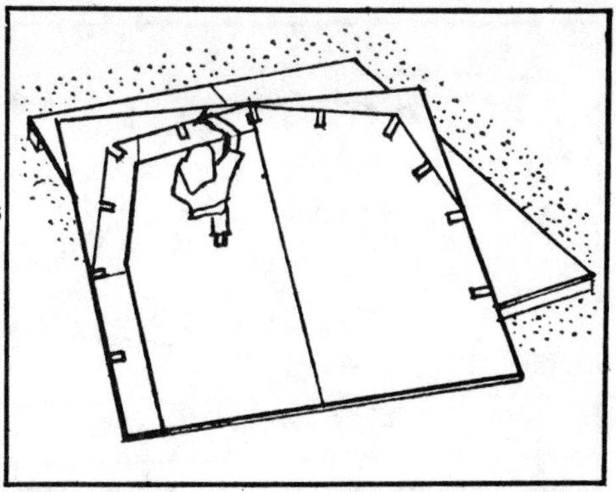

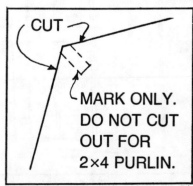

Fig. 17-8. Place template as shown and mark an outline on the back side of the siding.

Now the flooring is cut to a length of 7 feet 8 inches (for the 8-foot long building) and left at 4 feet wide. Nail the flooring in place on the runners, making sure the edges and sides are flush. See Figs. 17-1 through 17-5.

Now lay two picese of siding across a couple of sawhorses or on the floor, with 2 by 4s holding it off the floor surface. Butt these pieces of siding tightly. It might even be a good idea to lightly tack nail them once they're butted. Don't forget to use waterproof wood putty to fill the tack nail holes later. Place the template and mark the outline on the top sheet, making sure the back side of the siding is up. Cut along marked lines. Now the butted pieces have the door and frame parts tack nailed in place. At this point, the nailing must only be temporary. With the exception of the end rafter trim, this can be nailed if the fit is correct (Figs. 17-6 through 17-10).

Cutting and Nailing Parts

Now the saw is run up the door trim center and through the siding to make the door cutouts. You'll need to finish the corner cuts

Fig. 17-9. This step is very important.

CUT

MARK ONLY.
DO NOT CUT
OUT FOR
2×4 PURLIN.

Fig. 17-10. Cut siding on a marked line. Place 2 by 4s underneath siding so you won't cut into the floor. Repeat for the back wall.

with a handsaw for the neatest cuts. Now cut the center door slot, through trim and siding, again finishing with a handsaw. Once this is done, temporary braces are nailed as shown in Figs. 17-11 through 17-15. The hinges, hasp and swivel are screwed on.

The section is now turned over and rafter positions are marked. Rafters and studs are now nailed to the siding, using 8 penny nails. Then interior door trim is nailed in place, again with 8 penny nails.

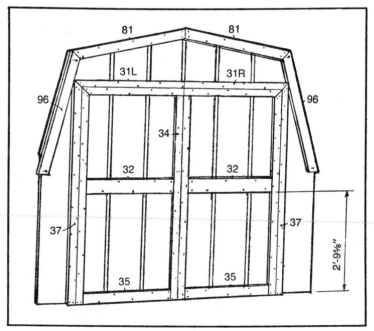

Fig. 17-11. Location of parts for the barn's exterior front section.

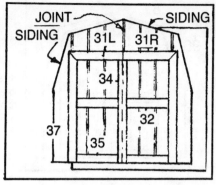

Fig. 17-12. Butt halves of siding and tack door and frame parts in place. Be sure the grooved line on part 34 is in line with the joint of the siding.

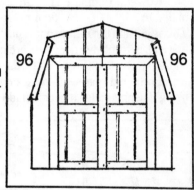

Fig. 17-13. Check position of all parts. Make sure part 96 will fit properly.

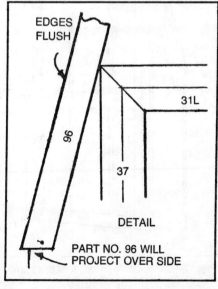

Fig. 17-14. Part 96 will project over the side.

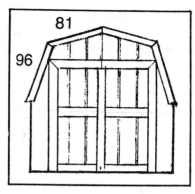

Fig. 17-15. Nail on trim, part 81. Keep edges flush. Check all parts for fit and nail with 6d nails. Do not nail on grooved lines. Do not complete nailing part 96.

Now the template is again used on butted siding panels, tack nailed to 2 by 4s on the floor or sawhorses, to mark the back wall. This is cut and the rafters and studs marked for position and nailed in place as in front. No door cuts are needed here so the job is simpler and a bit quicker (Figs. 17-16 through 17-31).

Front and rear sections are now braced and nailed to the floor using 8 penny nails. It's a good idea to remove the braces from the

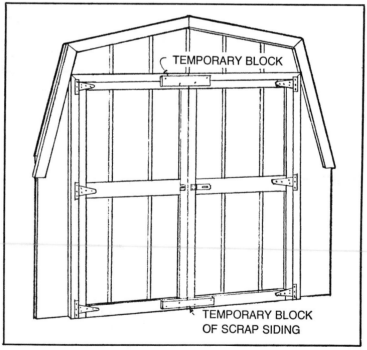

Fig. 17-16. Exterior front view of the barn.

344

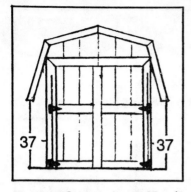

Fig. 17-17. Cut through parts 37 and the siding on grooved lines. Apply lower and middle hinges.

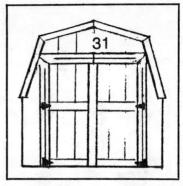

Fig. 17-18. Cut part 31. Finish cutting corners with the saw and apply upper hinges.

doors. The 16 penny nails are now used to nail all the purlins in place, with the vertical 2 by 4 studding nailed with 8 penny nails.

Siding and Roof Panels

Siding is now cut to size, in this case 4 feet wide, with a height of 3 feet 8 inches. Two pieces are needed for each side. The gauge blocks are used to place the siding, and it is nailed with 6 penny nails. The barn can still be twisted a bit, and you may need to give it a yank or two to get the siding pieces tight and flush with the end walls (Fig. 17-32 through 17-37).

Fig. 17-19. Hand-sawing corners.

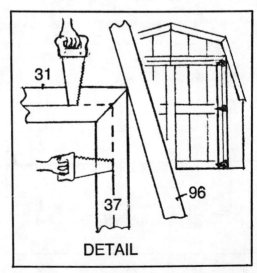

DETAIL

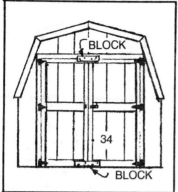

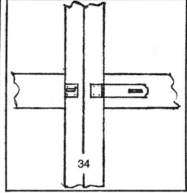

Fig. 17-20. Cut part 34 on grooved line. Finish with a handsaw. Nail on temporary blocks above and below to keep doors in place.

Fig. 17-21. Screw hasp and swivel on part 34. The hasp is on the right side; the swivel is on the left.

Roof panels are cut a full 8 feet long, with the top panel being 3 feet 6 inches wide and the bottom one 3 inches wider. Use two of each and apply the panels so the overhang is equal at the ends. Use 6 penny nails. As with any shingled roof, the first shingle course is laid upside down. Continue shingling as described in the general construction procedures earlier in the book.

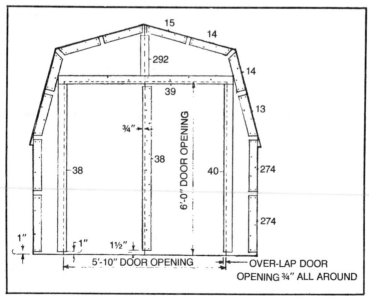

Fig. 17-22. View of the interior of the barn's front section.

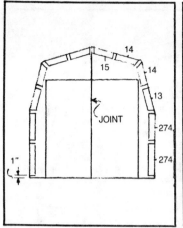

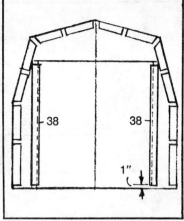

Fig. 17-23. Mark rafter positions on siding with a template.

Fig. 17-24. Nail part 38 to siding according to the dimensions in Fig. 17-22.

The result will be a good looking wood storage building of decent size, at a reasonable cost and with a long life expectancy. Enough material will probably be left over from the cutting of roofing and siding panels to provide you with a good number of shelves if you wish to customize your interior. Exterior finish is up to you, as is the choice of plywood siding styles (Figs. 17-38 through 17-49).

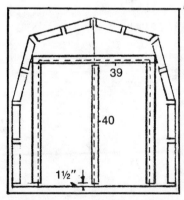

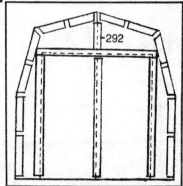

Fig. 17-25. Nail part 39 to siding. Use 8d nails. Also nail part 40 to the left-side door.

Fig. 17-26. Nail part 292. It must center on the joint of the siding. Use 8d nails. Turn the section over to complete nailing.

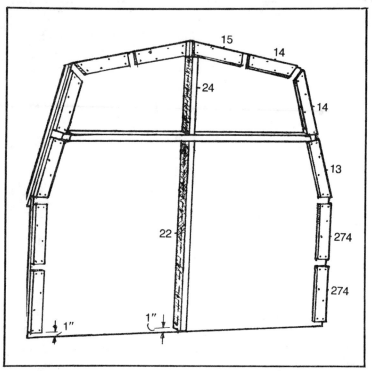

Fig. 17-27. View of the interior of the barn's back section.

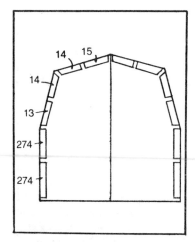

Fig. 17-28. Nail rafters to siding as on the front section. Keep edges flush. Use 6 penny nails.

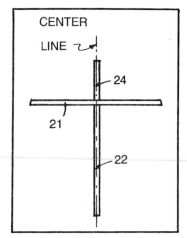

Fig. 17-29. Find the center of part 21. Nail part 22 to 21. Toenail part 24 to 21.

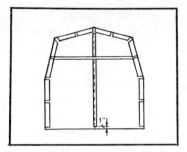

Fig. 17-30. Toenail frame to siding. Make sure centers of parts 22 and 24 are on the siding joint. Use 8d nails.

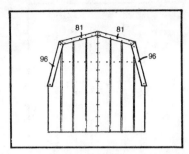

Fig. 17-31. Turn the section over and nail siding to the frame. Nail on trim as on the front section. Use 6 penny nails. Temporarily nail part 96 only.

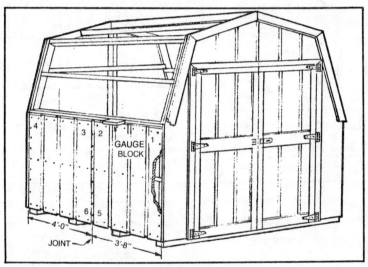

Fig. 17-32. Assembly detail for the barn.

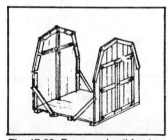

Fig. 17-33. Brace and nail front and back sections to the floor. Make the brace from scrap pieces of siding. Remove blocks from doors.

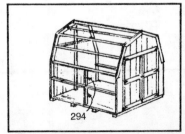

Fig. 17-34. Nail all 2 by 4 purlins in place. Use 16 penny nails. Nail vertical 2 by 4s and part 294 in place. Use 8d nails.

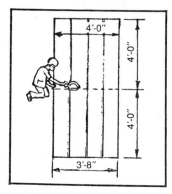

Fig. 17-35. Cut siding to size as shown, with two pieces for each side.

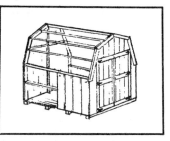

Fig. 17-36. Use gauge blocks to place the first siding panel. Use 6 penny nails. Repeat for the other side.

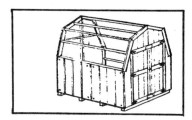

Fig. 17-37. Nail other siding panel in place at 3 and 4 in Fig. 17-32. Make sure the joint of the siding is tight and flush with end walls. Then nail at 5 and 6 in Fig. 17-32. Repeat for the other side.

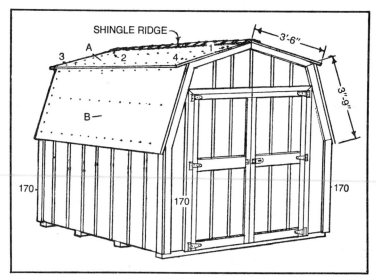

Fig. 17-38. Roof assembly detail.

350

Fig. 17-39. Lay panels with the groove side down. Cut two upper roof panels (A in Fig. 17-38) and two lower roof panels (B in Fig. 17-38).

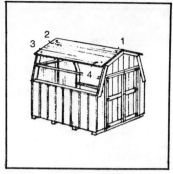

Fig. 17-40. Apply both top roof panels with equal overhang. Nail 1 and 2 as shown. Then nail 3 and 4.

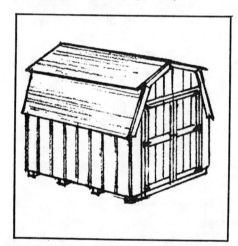

Fig. 17-41. Apply both lower roof panels.

Fig. 17-42. Nail asphalt shingles to the roof ridge and overlap to show about a 9-inch exposure. Use galvanized roofing nails.

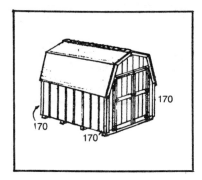

Fig. 17-43. Complete nailing of trim part 170 at all four corners with 6 penny nails.

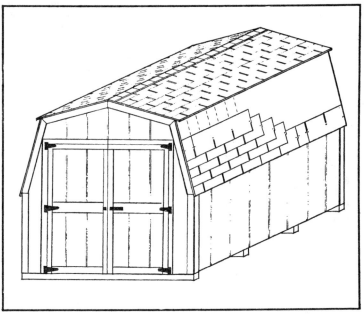

Fig. 17-44. View of the optional shingles.

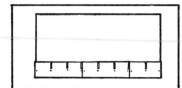

Fig. 17-45. Nail the first shingle up-side down, extending ¼-inch past the left corner and bottom. Continue shingling to the right and cut off with a ¼-inch overhang. Use roofing nails.

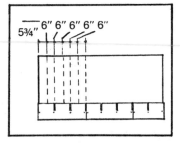

Fig. 17-46. Measure and mark lines on the roof panel as shown.

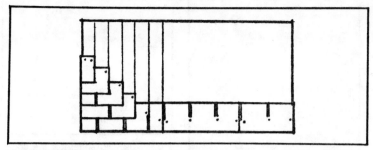

Fig. 17-47. Cut 6 inches off one shingle and nail on top of the row that is nailed upside down. On the second row start the first shingle with 12 inches cut off. Start each succeeding row with an additional 6 inches cut off on the first shingle.

Fig. 17-48. Finish nailing shingles on the roof panel. Trim off the right edge of the panel with a knife, leaving a ¼-inch overhang. Complete the process for the other three roof panels.

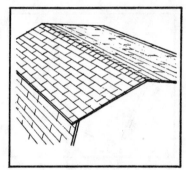

Fig. 17-49. Cap shingles can be obtained from a dealer. Or you can make them by cutting shingles on center of grooves. Nail on cap shingles with 5 inches exposed.

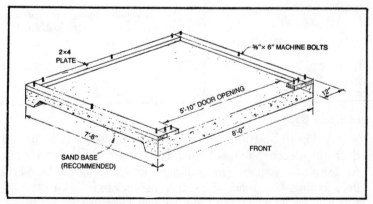

Fig. 17-50. Concrete floor section.

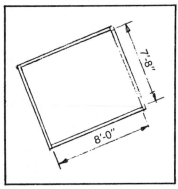

Fig. 17-51. Cut forms to size and nail them together.

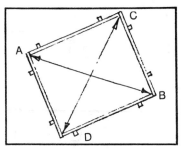

Fig. 17-52. Dig space for the concrete (4 inches deep and deeper at the edge). Place the form in a hole and stake one corner of the form. To square the form, measure diagonals and make AB and CD equal. Set the rest of the stakes.

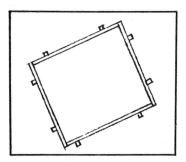

Fig. 17-53. Level forms, nail to the rest of the stakes and pour concrete.

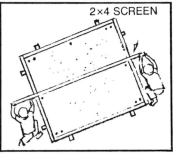

Fig. 17-54. Screed concrete. Use a long, straight 2 by 4. Place ⅜ by 6-inch machine bolts 1¾ inches from the edge. Finish troweling the slab. Let the concrete cure.

Fig. 17-55. Remove forms. Cut 2 by 4 plates to size. Bolt to the slab as shown.

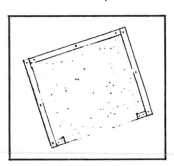

Concrete Base

As Figs. 17-50 through 17-55) show, it is also possible to build the barn on a concrete base. I would recommend going down below frost depth for your area and planting at least an 8-inch wide, 6-inch deep footing. Use 4-inch wide concrete block to build up a foundation wall inside which you can then pour your slab.

Chapter 18
Projects

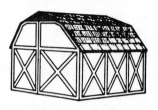

While imagination will provide you with some projects to work on around your home, a bit of work and thought may be saved if some of the plans in this chapter can be adapted. Several are set to be done exactly as indicated in the illustrations. Others can be added to, subtracted from, or otherwise changed to suit a variety of uses.

GAZEBO

This is just about the most complex of all our plans. The project can provide a shady and dry recreation area in almost any yard with a fairly level spot—at least 10 by 14 feet in size. All materials are pressure-treated. The gazebo shown was built of Wolmanized lumber and plywood for durability. You will need four sheets of 4 by 8-foot ½-inch exterior grade plywood and a single sheet of ¾-inch exterior plywood. Add to that 28 2 by 4s 8 feet long and eight 2 by 6s 12 feet long, plus two more 8 feet long. In addition, you need 24 machine bolts, 5½ by ⅜ inches; eight bolts, 4 by ⅜ inches; and 12 bolts, 3 by ⅜ inches. Eighty-eight washers are required for the machine bolts. Also needed are eight flush style joist hangers, two rolls of 15-pound asphalt roofing paper, 2 pounds of 10 penny galvanized nails, 2 pounds of ¾-inch galvanized roofing nails, and 1½ squares of cedar shingles.

The site must be cleared, if needed, and should be about 10 by 14 and reasonably level. Prepare the location before construction starts, because the unit is built in place. Position the gazebo as in Fig. 18-1 by working out an 8 by 11-foot rectangle around what will be the center line of the gazebo. At each of the four corners, you must dig a hole 2 feet in diameter and 2 feet deep. Actually the holes must go slightly lower and be lined with rock or brick, as in Fig. 18-2.

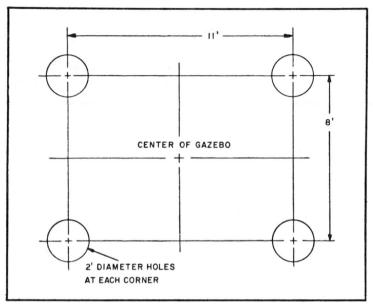

Fig. 18-1. Position the gazebo by establishing an 8 by 11-foot rectangle around its center line. At each of the four corners of the rectangle, dig a hole 2 feet in diameter and 2 feet deep.

Benches

Once the site is prepared, the benches are begun. Cut 24 2 by 4s to a length of 7 feet 4½ inches. These form the seats and back slats. Following Fig. 18-3, draw the patterns for the four bench ends on the sheet of ¾-inch exterior plywood and cut

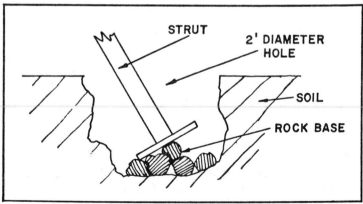

Fig. 18-2. To assure a firm base in soft soil, dig the holes slightly deeper and line the bottoms with stones or bricks.

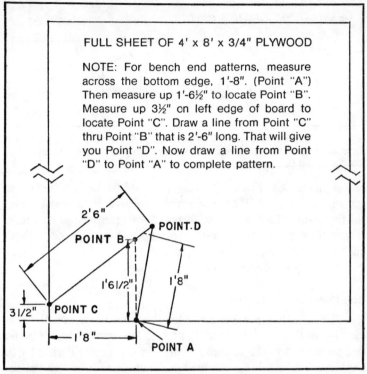

FULL SHEET OF 4' x 8' x 3/4" PLYWOOD

NOTE: For bench end patterns, measure across the bottom edge, 1'-8". (Point "A") Then measure up 1'-6½" to locate Point "B". Measure up 3½" on left edge of board to locate Point "C". Draw a line from Point "C" thru Point "B" that is 2'-6" long. That will give you Point "D". Now draw a line from Point "D" to Point "A" to complete pattern.

2'6"

POINT B

POINT D

1'6 1/2"

1'8"

POINT C

3 1/2"

1'8"

POINT A

Fig. 18-3. Draw the patterns for four bench ends on the ¾-inch exterior plywood and cut the four end pieces out.

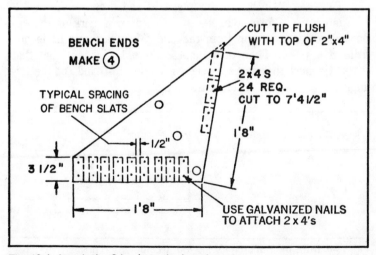

BENCH ENDS MAKE ④

CUT TIP FLUSH WITH TOP OF 2"x4"

2x4S 24 REQ. CUT TO 7'4 1/2"

TYPICAL SPACING OF BENCH SLATS

1/2"

1'8"

3 1/2"

1'8"

USE GALVANIZED NAILS TO ATTACH 2 x 4's

Fig. 18-4. Attach the 2 by 4s to the bench ends.

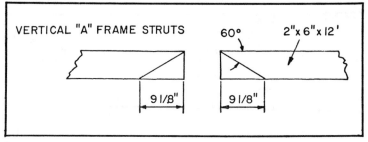

Fig. 18-5. At one end of each 2 by 6, measure 9⅛ inches along one edge from the corner and draw a diagonal line from that point to the opposite corner of the board.

those pieces out. The 2 by 4s are attached to the bench ends as shown in Fig. 18-4. Please note that the back slats are attached so that 4-inch (nominal) face forms the surface of the backrest. The nine 2 by 4s used for the seat bottom are spaced with ½ inch between them, as are the three used for the back. The second bench goes together in exactly the same way.

Top Plates and Crossarms

The benches are set aside for the time being and the fabrication of the vertical struts is begun. The four 12-foot 2 by 6s are used to form the A-frame struts. At one end of each 2 by 6, measure 9⅛ inches up the lumber. Then draw a line to the diagonally opposite corner, as shown in Fig. 18-5. This is cut off to form a 60 degree angle.

Take the ¾-inch exterior plywood and draw marks at 12-inch intervals to form four pieces. Then draw a mark on the opposite side 6 inches in from the end. Make a mark 12 inches from that point. Then draw another one 12 inches from that mark. Connect the marks as shown in Fig. 18-6 and cut the triangles out. These will serve as top plates.

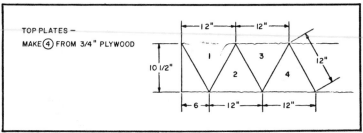

Fig. 18-6. Draw the patterns for the four triangular top plates.

358

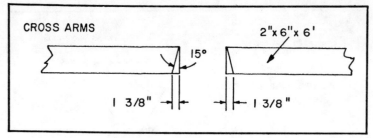

Fig. 18-7. Here's how to make the crossarms from the eight 6-foot long 2 by 6s.

The next step is to select the eight 6-foot long 2 by 6s to be used as the two double crossarms. A 15 degree cut is made at one end of each 6-foot length. To easily make the cut, measure up one edge 1⅜ inches and draw a diagonal line to the corner of the opposite edge (Fig. 18-7).

Vertical Assemblies

The 2 by 6 scrap even comes in handy. You'll need to cut spacers for use in the crossarms. These are cut to 4-inch length from scrap 2 by 6, with four required. Get two 12-inch spacers for the centers of the crossarms from other 2 by 6 scrap pieces. These spacers provide needed separation for the A-frame vertical struts, while the 12-inch spacers are used in bolting the

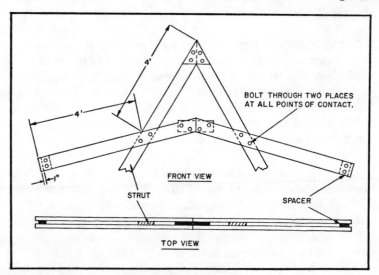

Fig. 18-8. Spacers are attached in this manner.

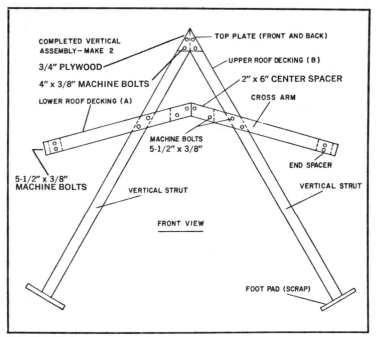

Fig. 18-9. Holes are to be drilled with a 7/16-inch wood bit.

halves of the crossarms together. The A-frame assembly is constructed by laying a set of crossarms on the ground, with the spacers attached as shown in Fig. 18-8. Measure in 4 feet from the ends of the crossarms and draw a line. Now lay two of the A-frame vertical struts on the crossarms. Measure 4 feet from the top (where the 60 detree angles meet) of the struts and draw another line. The crossarms and vertical struts join, as you see in Fig. 18-8, where the lines are drawn. Next you fasten the top plates to the apex of the vertical struts. Place the outer crossarm in position and nail it to the spacers and struts.

Figure 18-9 shows where the holes are to be drilled, using a 7/16-inch wood bit. This allows sufficient clearance for the ⅜-inch machine bolts to be used throughout the structure. Use 4-inch bolts to attach the top plates and 5½-inch bolts for the crossarms. Once you've got the vertical assembly, nail on 8-inch pieces of 2 by 6 as footpads. That gets you through the first A-frame, and the second goes together exactly the same way.

Once both vertical assemblies are completed, they can be stood in their holes and tripod-braced with 2 by 4s (Fig. 18-10).

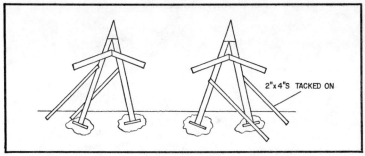

Fig. 18-10. Verticals can be supported by tacking 2 by 4s to them in a tripod fashion.

One of the completed benches is now placed between the vertical assemblies. The bottom of the bench should measure 13 inches from ground level. Use bracing, as shown in Fig. 18-11, to keep the bench at this height. Tack one end of the bench into place. You need a level to position the other end so the bench surface, or seat, is level. Now tack the second end of the bench in place. Drill through the vertical struts and bench ends, again with a 7/16-inch bit, and use 3 inch machine bolts for fastening. All machine bolts should have washers on both sides (Fig. 18-12).

Leveling

Do the second bench. Once that is in place, lay a straight 2 by 4 across the two benches and place your level on that. If the gazebo isn't leveled, now is the time to do it. Place stones under foot pads on the low end to get it level. Once the gazebo

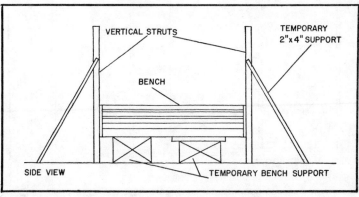

Fig. 18-11. Place temporary supports under the bench.

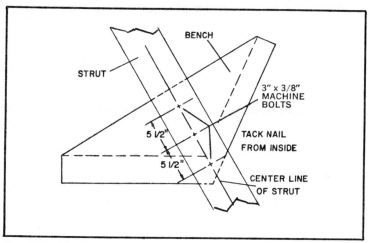

Fig. 18-12. Machine bolts should have washers on both sides.

is leveled, the soil is replaced in the holes and tamped down well. The structure should now be stable and secure.

Joist hangers are nailed in place and the joists are nailed (Fig. 18-13). End joists are 2 by 6 and inner joists are 2 by 4s, spaced as you see in Fig. 18-13. Joist hangers by Teco can be found in various sizes to ease the job and make metal cutting unnecessary.

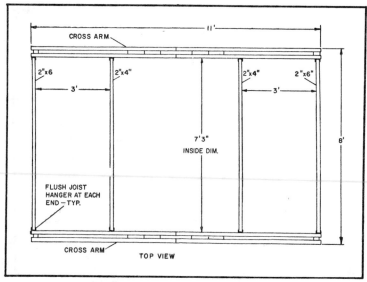

Fig. 18-13. Nail joist hangers in place on the crossarms.

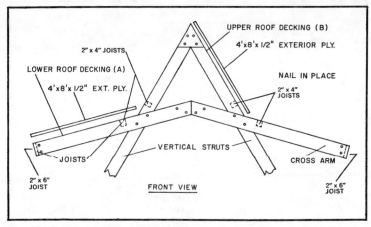

Fig. 18-14. Nail the plywood in place.

Roof Sheathing and Shingles

The lower roof sheathing of ½-inch exterior plywood can now be placed. It is attached to the crossarms and joists with 6 penny nails (Fig. 18-14). This sheathing should easily support you so that you can set the upper sheathing in place using 6 penny nails. Roofing paper is rolled out over the sheathing and stapled in place. Use a double thickness over all joints (Fig. 18-15).

The first course of shingles is placed loose as a trial layer, with the thick edge to the eave. It is positioned along the left edge to serve as a spacing guide (Fig. 18-16). The actual first

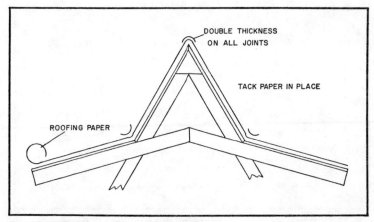

Fig. 18-15. Staple roofing paper in place.

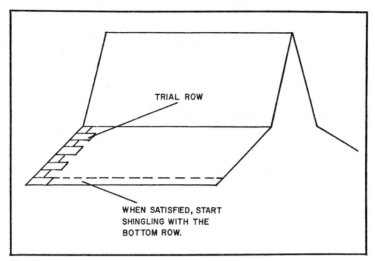

Fig. 18-16. Place a trial layer of shingles over the roofing paper, along the left edge.

course goes along the bottom edge of the eave. Seven courses are recommended by Koppers Company. For the upper portion of the roof, start with a trial course one more time. Mark each shingle with chalk, as the shingles are unlikely to stay in place on this steep a pitched roof. The cap row, after all shingles have

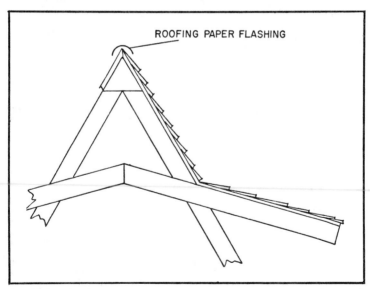

Fig. 18-17. When shingling is complete, cap with asphalt flashing.

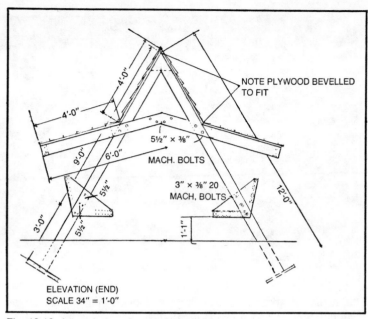

NOTE PLYWOOD BEVELLED
TO FIT

4'-0"

4'-0"

9'-0"

6'-0"

5½" × ⅜"
MACH. BOLTS

3" × ⅜" 20
MACH. BOLTS

5½"

3'-0"

5½"

5½"

12'-0"

1'-1"

ELEVATION (END)
SCALE ¾" = 1'-0"

Fig. 18-18. An end elevation of the gazebo.

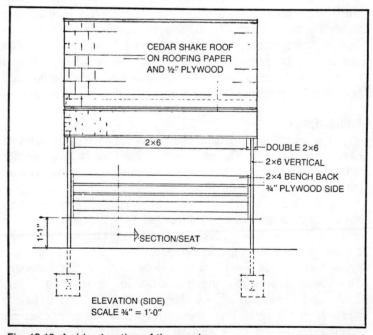

CEDAR SHAKE ROOF
ON ROOFING PAPER
AND ½" PLYWOOD

2×6

DOUBLE 2×6
2×6 VERTICAL
2×4 BENCH BACK
¾" PLYWOOD SIDE

1'-1"

SECTION/SEAT

ELEVATION (SIDE)
SCALE ¾" = 1'-0"

Fig. 18-19. A side elevation of the gazebo.

Fig. 18-20. The completed gazebo.

been placed and nailed, can be made of roofing paper. Or you can buy a metal ridge cap (Fig. 18-17).

Figures 18-18 and 18-19 provide elevations of the gazebo, while Fig. 18-20 shows the completed gazebo in use. The finished gazebo is open to the breeze, yet shaded and dry (except in driving rains). It provides room for 10 to 12 people. The gazebo can be stained, painted or left to weather with few maintenance worries because of the cedar shingles, galvanized fasteners and pressure-treated wood.

PICNIC TABLE

A picnic table with attached benches can provide a lot of convenience and fun. The one we describe here is exceptionally sturdy and made of pressure-treated wood. It should provide many years of trouble and maintenance-free use (Fig. 18-21).

The table is 6 feet long, but can be extended easily if you have a large family. Simply replace the 6-foot lengths of 2 by 6 in Table 18-1 with 8-foot lengths. For longer tables, it would be a good idea to add a third leg support. The table can usually get by without it, but the benches will need it—at the midpoint. This should easily allow sturdy construction up to 12 feet long. Table 18-1 lists materials for the 6-foot table. Tools include a wrench, brace and bit, saw, square, hammer and some sandpaper. The job does go more quickly with an electric saw and drill.

Fig. 18-21. Diagram of the picnic table.

The end braces are constructed first, as they provide both the table legs and the seat supports. Table legs are 2 by 6s cut 34 inches long, with a 3¼-inch diagonal cut parallel to each end to accept the slant of the table legs. The cross braces are again 2 by 6 lumber, cut 58½ inches long for the lower, or seat, brace and 29½ inches long for the top brace. Once the cuts are made, lay the pieces on the ground so the table legs are flush with the top edge of the table top brace and placed 4¾ inches from its ends (Fig. 18-22). Set the seat brace so the seat edge is 14 inches above the bottom of the table legs. Carefully drill the rquired ⅜-inch holes through the cross braces and table legs. I would start at the top and drill one hole through a cross brace. Place a loosely fitted bolt, drill the next hole and then go to the seat brace. Asemble the two end supports using the bolts. Washers should be under the nut end if carriage bolts are used and under both ends if machine bolts are used.

Table 18-1. Materials List for the Picnic Table.

9 PCS. 2×6—6' LONG	12 GALVANIZED BOLTS, ⅜"×3½" LONG
4 PCS. 2×6—3' LONG	12 ⅜" STANDARD WASHERS
1 PC 2×6—10' LONG	8d GALVANIZED NAILS
2 PCS. 2×4—10' LONG	

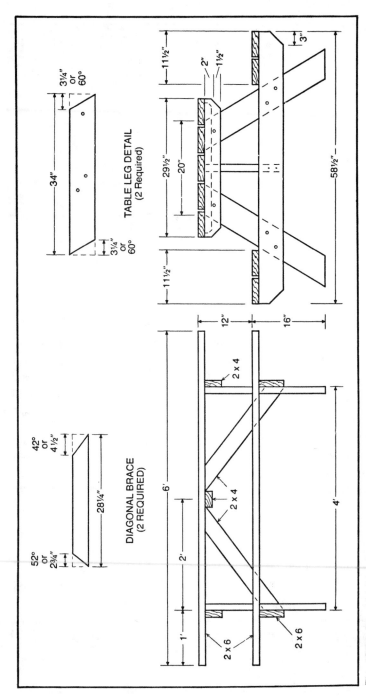

Fig. 18-22. Picnic table plans.

Fig. 18-23. Getting the leg assemblies in place.

Either brace the end assemblies using scrap lumber and tack nailing so they stand 4 feet apart, or get a helper to hold them on the marks (Fig. 18-23). Nail the first outside table top 2 by 6 flush with the outside edges of the leg assembly top brace, keeping the bark side up to minimize cupping. Place and

Fig. 18-24. Final assembly of the table.

Fig. 18-25. The table is ready for use.

nail the second outside board in the same manner. Go on to nail the rest of the boards in place as in Fig. 18-24.

To get an extra sturdy table, place diagonal cross braces, cut as shown in Fig. 18-22, under the table and carefully nail. Once this is done, you have a table like the one in Fig. 18-25. It will last for many years.

SHED ADDITIONS

Sturdy shed additions to homes or other outbuildings can be made using dimension lumber and Teco framing anchors. If the shed is added to an outbuilding and needs no floor, use pressure-treated lumber at the corners. Sink the 4-inch square posts below ground level, setting them in concrete. This will provide a sturdy base for the rest of the shed. Use pressure-treated lumber again for the sole plates, and at least a 2 by 6 ledger board bolted to the studs in the house or outbuilding wall. Figure 18-26 gives framing details using the anchors, while Fig. 18-27 shows some of the possible uses for such a shed. If the shed is built with an open side, face that side away from prevailing winds in your area. Use 4 by 4 pressure-treated lumber at all corners.

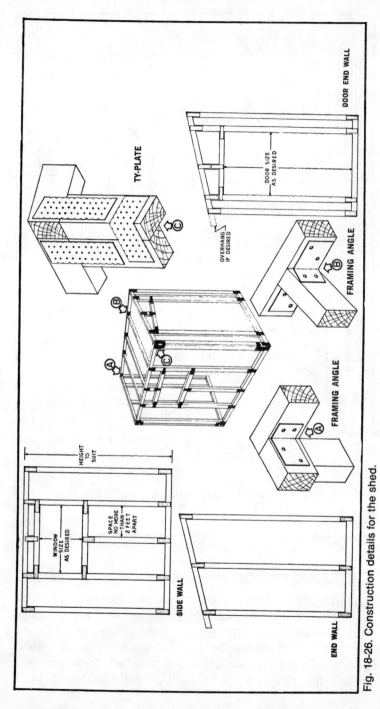

Fig. 18-26. Construction details for the shed.

371

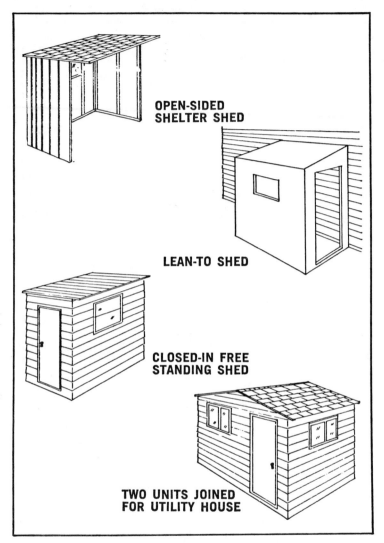

OPEN-SIDED SHELTER SHED

LEAN-TO SHED

CLOSED-IN FREE STANDING SHED

TWO UNITS JOINED FOR UTILITY HOUSE

Fig. 18-27. Variations on the shed theme.

LANDSCAPING TIMBERS

Landscaping timbers are always of pressure-treated lumber, often treated with creosote. Many timbers are being treated with waterborne preservatives which are cleaner to handle and work with. Timbers can be used for many things—tree well, terracing for gardens, sandboxes, compost bins, steps and other items (Figs. 18-28 through 18-34). Figures 18-35 through 18-43

Fig. 18-28. The tree well.

will provide more ideas, as well as a look at the different kinds of framing anchors made to fit landscaping timbers. It is also possible to spike the timbers together, though the Teco anchors are generally easier and quicker to use.

I am going to install a brick walk and gravel drive in front of my house this summer. Originally I had planned to use a soldier course of bricks to hold the main walk in place. The use of landscaping timbers as a border for the brick strikes me as a very practical alternative and a great time saver. It will also allow me to use those 1000 oversized brick to greater advantage

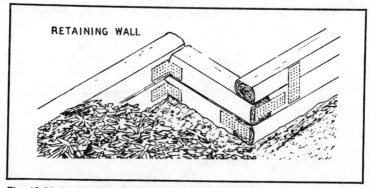

Fig. 18-29. Landscaping timbers can form a retaining wall.

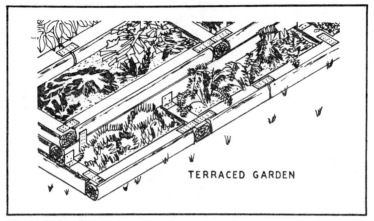

Fig. 18-30. Use the timbers for a terraced garden.

over a wider area in front of the porch, since only the large sides of the brick need be used. I may just continue the timbers out along the edge of the T driveway shape and fill in with gravel to the tops of the half-buried timbers. Costs should be cut, at least for the walk, as the soldier course of brick doesn't cover much space.

Later on I can probably save time by erecting a retaining wall of timbers or even poles several feet down the hill. The possibilities may not be endless with landscaping timbers, but there are certainly quite a few to choose from.

BARN STORAGE SHED

We've already looked at the kit barn from Jer Manufacturing, but Teco provides us with a chance to start from scratch.

Fig. 18-31. Make a sandbox out of timbers.

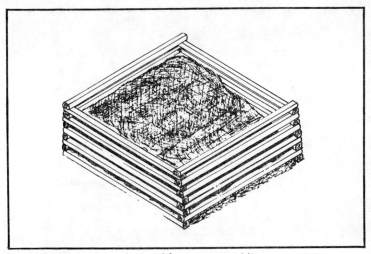

Fig. 18-32. Timbers can be used for a compost bin.

You can use their angle cuts and templates to form your own gambrel roof barn. As Fig. 18-44 shows, it can be done in several ways, including setup for use as a greenhouse. Make sure to install ventilators, as covered in the chapter on greenhouses.

The Teco barn is designed for an 8-foot width and an 8-foot length. I see no difficulty at all in extending the barn by 4 to 8 feet if you need more space.

18-33. Another use of timbers is for a planting bed.

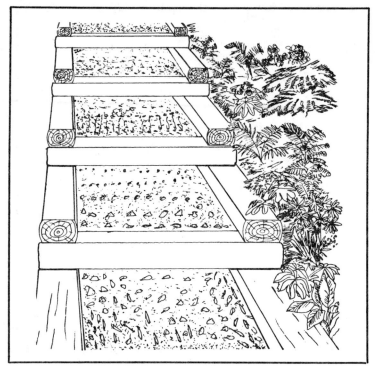

Fig. 18-34. Timbers can be used when building steps.

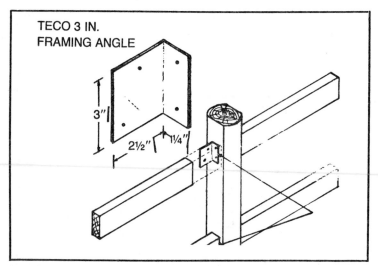

Fig. 18-35. Teco framing angles can fasten dimension lumber rails to land-scape timber fence posts.

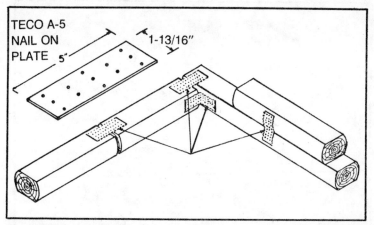

Fig. 18-36. Teco A-5 plates are utilized in end splicing and stack tying landscaping timbers.

Siting and layout is done as already explained in the general construction chapter, making sure the footings or piers meet local needs. It would also be a good idea to check the barn features with your local building department if their rules tend to be strict. See Table 18-2 for materials.

You start by cutting framing pieces with the templates (Fig. 18-45). Assemble as shown in the intermediate drawings (Figs. 18-46 through 18-52). Cut the first framing piece with one end

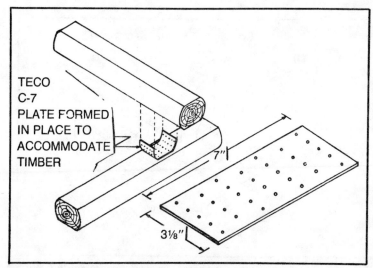

Fig. 18-37. Teco C-7 plates are handy devices.

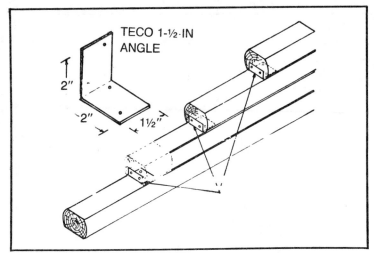

Fig. 18-38. Step-stacked landscaping timbers end fastened with either 3-inch or 1½-inch framing angles.

at the indicated 18 degree angle. Then measure to 2 feet 7½ inches and cut that end to 18 degrees. You can now cut three more pieces to this pattern. Cut the next piece at 18 degrees on only one end, and have the long side 4 feet in length. Two of these are needed for each frame, and you need a total of four of these frames for an 8-foot long building. Add another frame for every 32 inches of added depth. These frames are joined, with

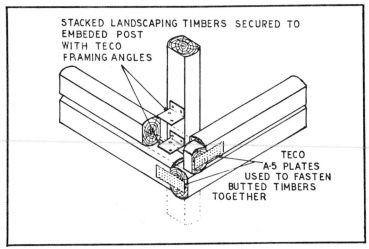

Fig. 18-39. Teco A-5 plates are used here to fasten butted timbers together.

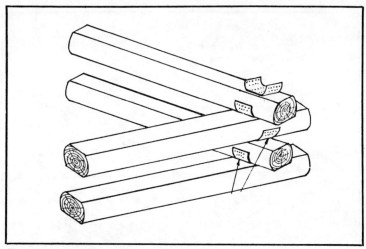

Fig. 18-40. Teco C-7 plates connect a split-rail type structure.

nailing of the plates done on a flat surface, and then erected onto the sill plates. Purlins are installed and the building is sided and roofed.

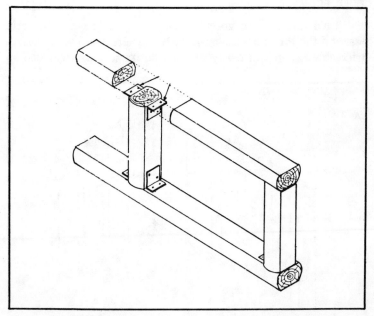

Fig. 18-41. Teco framing angles are utilized to fasten landscaping timbers together at right angles.

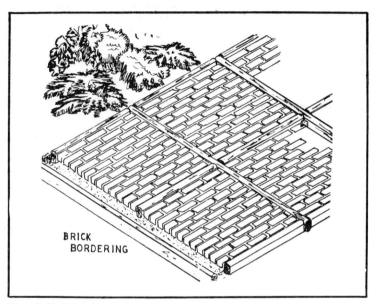

Fig. 18-42. Brick bordering.

COLD FRAME

Cold frames, like greenhouses, can give you a good early start on the growing season no matter where you live. An electric cable heater can easily turn a cold frame into a hot bed so

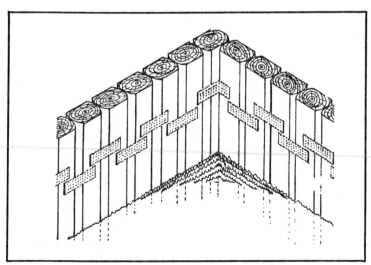

Fig. 18-43. Raised planting bed.

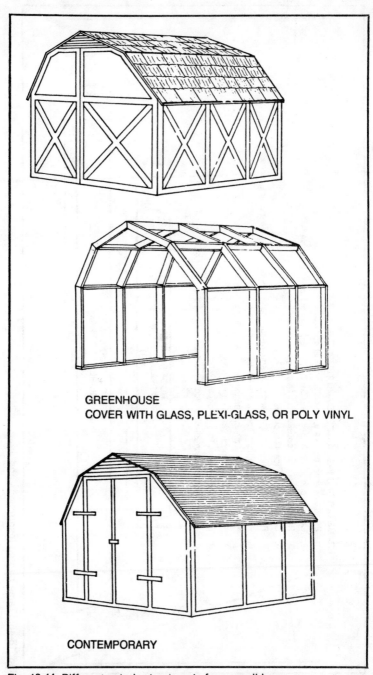

GREENHOUSE
COVER WITH GLASS, PLEXI-GLASS, OR POLY VINYL

CONTEMPORARY

Fig. 18-44. Different exterior treatments for a small barn.

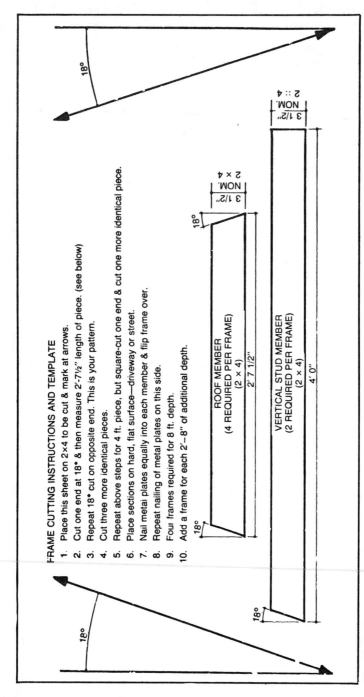

FRAME CUTTING INSTRUCTIONS AND TEMPLATE

1. Place this sheet on 2×4 to be cut & mark at arrows.
2. Cut one end at 18° & then measure 2′-7½″ length of piece. (see below)
3. Repeat 18° cut on opposite end. This is your pattern.
4. Cut three more identical pieces.
5. Repeat above steps for 4 ft. piece, but square-cut one end & cut one more identical piece.
6. Place sections on hard, flat surface—driveway or street.
7. Nail metal plates equally into each member & flip frame over.
8. Repeat nailing of metal plates on this side.
9. Four frames required for 8 ft. depth.
10. Add a frame for each 2–8″ of additional depth.

ROOF MEMBER
(4 REQUIRED PER FRAME)
(2 × 4)
2′ 7 1/2″

VERTICAL STUD MEMBER
(2 REQUIRED PER FRAME)
(2 × 4)
4′ 0″

2 × 4 NOM. 3 1/2″

2 × 4 NOM. 3 1/2″

Fig. 18-45. Frame cutting insturctions and template.

Table 18-2. Materials for the Barn Storage Shed.

QUAN.	DESCRIPTION	QUAN.	DESCRIPTION
28	2″ × 4″ × 8FT. LONG	40	TECO C-7 PLTS.
9	4′ × 8′ × ½″ PLYW'D	30	TECO JOIST HGR.
2	1″ × 4″ × 6 FT. LONG	12	TECO ANGLES
1 ROLL	ROOFING FELT	30	TECO A-5 PLTS.
1 GAL	ROOF. CEMENT	3	3 BUTT HINGES
1 GAL	BARN-RED PAINT	1	HASP & LOCK
5 #	6d COM. NAILS	10 BG.	90# CONC. MIX
2 #	12d COM. NAILS		OR
2 #	½″ ROOF. NAILS	4	6″ × 8″ × 8′ RAIL TIE

that plants can be started long before the frost is out of the ground (Fig. 18-53). The cold frame is easily assembled using framing anchors. If it is on or in the ground, be certain the framing anchors are galvanized and the wood used is pressure-treated. Use fiber glass sheeting or corrugated fiber glass for the cover for longest cold frame life. Polyethylene sheet is okay, but it will often pull loose under the weight of water or

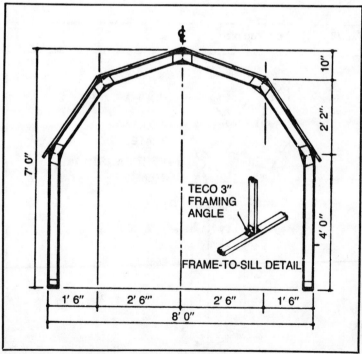

Fig. 18-46. Frame layout for the barn storage shed.

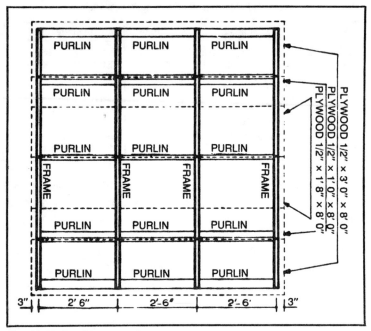

Fig. 18-47. Roof framing plan.

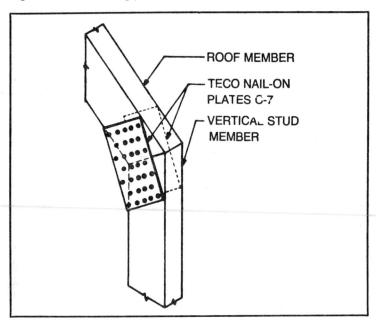

Fig. 18-48. Frame joint detail.

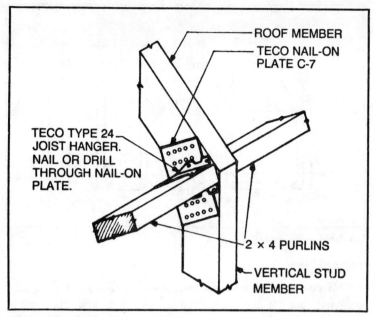

Fig. 18-49. Purlin attachment detail.

snow if you don't want to store the cold frame. Use 2 by 2s for the frame and nominal 1-inch lumber for sides, front and back in the sizes you need. Frame with the proper anchors (Fig. 18-54).

You may have some ideas for projects of your own. A look at some ideas for deck decoration may give you a few more.

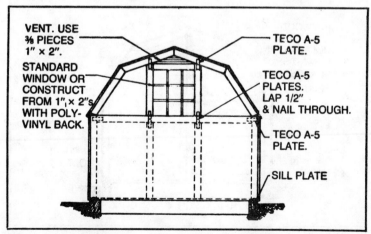

Fig. 18-50. Rear elevation.

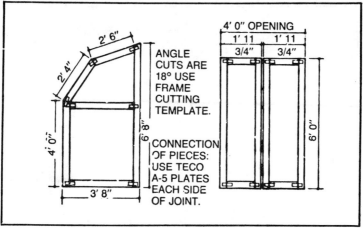

Fig. 18-51. Door details.

DECK DECOR

We've covered the construction of planters and benches for deck use, but not everything that goes on a deck will be built at home. And not all of it needs to be bought. If you don't feel like building a planter or two this year and would still like some deck greenery, terra-cotta planters and pots are readily available, and blend well with decks (Figs. 18-55 and 18-56). The variety of plants for the deck is wide. You might consider some dwarf fruit trees.

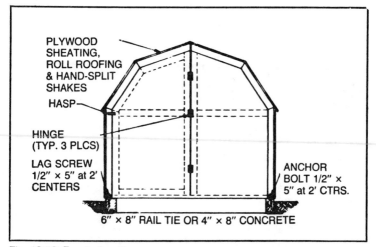

Fig. 18-52. Front elevation.

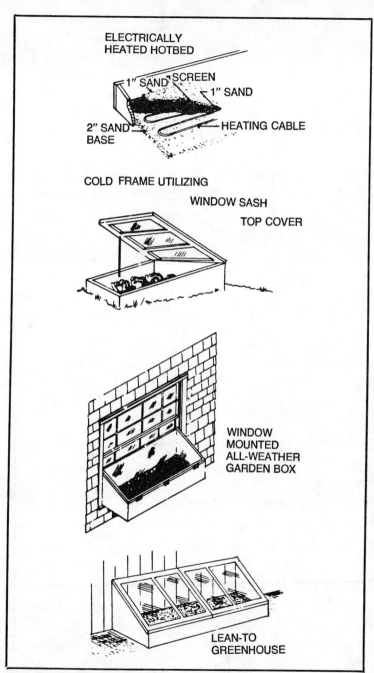

ELECTRICALLY
HEATED HOTBED

1" SAND SCREEN
1" SAND
2" SAND BASE
HEATING CABLE

COLD FRAME UTILIZING
WINDOW SASH
TOP COVER

WINDOW
MOUNTED
ALL-WEATHER
GARDEN BOX

LEAN-TO
GREENHOUSE

Fig. 18-53. Cold frame variations.

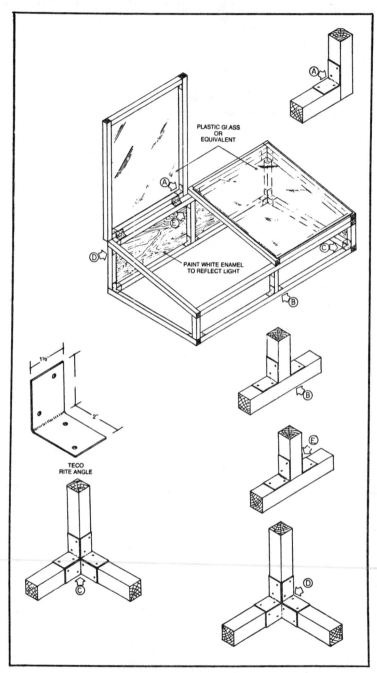

Fig. 18-54. Construction details for the cold frame.

Fig. 18-55. Plants do well as deck decorations.

Fig. 18-56. Shrubs or trees can be planted in larger containers for decoration.

Fig. 18-57. Cable spools can be used as tables.

Deck furniture is best selected for its weather resistance in addition to attractiveness. There is nothing more annoying than having to move furniture indoors the minute a few drops of rain fall.

Fig. 18-58. A deck can be a garden place, constructed and furnished as your imagininaition and wallet allow.

Old nail kegs can be set up with a round top and covered with a variety of materials or painted to make handy tables. Telephone cable or wire spools can serve the same purpose (Fig. 18-57). Decks are informal, so there is little harm in picking up a cast-off item and turning it into an attractive addition to your outdoor setting (Fig. 18-58). Set it up to please yourself—this holds true for all projects in this book.

Glossary

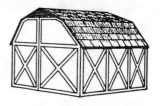

air-dried lumber: Lumber that has been dried in yards or sheds for a length of time. Minimum moisture content of 12 percent is considered best, but in humid areas 19 percent is acceptable.

anchor bolts: Bolts used to secure a wooden sill plate to a concrete or masonry floor or wall.

asphalt: A tarlike variety bitumen, found in a natural state or obtained by evaporating pertroleum. It is insoluble in water, but soluble in gasoline and will melt when heated. Asphalt is used for many purposes in the building industry.

backfill: Replacement of excavated earth in the trench around or against a basement foundation or behind a retaining wall.

balusters: Small vertical members in a railing used between a top rail and stair treads or a bottom rail.

balustrade: A railing made up of balusters, top rail and often a bottom rail, used on the edges of stairs, balconies and porches.

base or baseboard: A board placed against the wall and next to the floor to finish properly between floor and wall.

base molding: Molding used to trim the upper edge of baseboard.

base shoe: Molding used next to the floor on interior baseboard.

batten: Narrow strips of wood used to cover joints.

batter board: One of a pair of horizontal boards nailed to posts set at the corners of an excavation. It is used to indicate the desired level and as a point for fastening mason's cord to show building outlines.

beam: A structural member that supports a transverse load.

bearing partition: Any partition supporting a vertical load as well as its own weight.

bearing wall: Any wall that supports a vertical load in addition to its own weight.

blind nailing: Nailing in a manner that keeps the nail heads from showing on the face of the work; used in shingling and to nail tongue and groove flooring most of the time.

brace: An inclinded piece of lumber applied to a wall or board to stiffen the structure, often temporary and sometimes permanent, as in corner bracing of structures.

brick veneer: Facing of brick laid against, or close to, the sheathing of a frame wall and fastened to the sheathing.

bridging: Small wood or metal pieces inserted in a diagonal position between floor joists near midspan to act in spreading the action of loads.

built-up roof: A roof composed of three to five layers of asphalt felt laminated with hot asphalt or tar, and finished with gravel or crushed slag. It is used most often for flat or low pitched roofs.

butt joint: The junction where two ends of the members meet in a square cut joint.

casing: Molding used to trim door and window openings at the jambs.

collar beam: Nominal 1 or 2-inch pieces of lumber connecting opposite roof rafters to help stiffen the roof.

column: Architecturally a perpendicular supporting member, consisting of a base, shaft and capital or cap.

conduit (electrical): Pipe, most often metal, through which wire is passed. Used to protect against mechanical damage.

corner braces: Diagonal braces at the corners of frame structures used to stiffen the wall. A let-in brace is set into notched studs, while a cut-in brace is made of nominal 2-inch lumber, cut and set in between studs diagonally.

crawl space: A shallow space below the structure, normally enclosed by the foundation wall.

d: See *penny*.

decay: Disintegration of wood or other materials, often through the action of fungi.

direct nailing: Nailing perpendicular to the initial surface or junction of the two pieces being joined. The same as face nailing.

downspout: A pipe, usually metal and most often aluminum, to carry rainwater from gutters.

dry wall: Interior wall covering material such as gypsum wallboard, plywood or other forms of paneling, usually in large sheets.

eaves: The lower edge of the roof as it projects over the wall.

fascia: A flat board used to cover the ends of rafters at the eaves, often used in combination with moldings.

footing: A masonry section, most often concrete, in a rectangular shape wider than the bottom of the foundation wall or pier it supports.

foundation: The supporting part of a structure under the first floor structure, often below grade. It includes the footings.

frost line: Depth of frost penetration in ground. The depth varies according to area. All footings must be placed below frost line to prevent heaving.

fungi: Microscopic plants that live in damp wood causing mold, stain and rot.

fungicide: Chemical poisonous to fungi.

furring: Strips of wood applied to a wall or other surface to even it and serve as a base for fastening finish materials.

gable: The part of the roof above the eave line of a double sloped roof.

gable end: An end wall with a gable.

girder: A large beam of wood or steel used to support heavy, concentrated loads at particular points along its length.

grout: Mortar mixed to a consistency that allows it to flow into joints and cavities of masonry work to solidly fill them.

gusset: A flat wood, plywood or metal member used to provide a strong connection where pieces intersect. It is often used at joints of wood roof trusses and fastened with nails, screws, bolts or glues.

gutter: A shallow channel of metal set below and along the eaves to carry off rain water from the roof surface.

header: A beam placed perpendicular to joists, to which the joists are nailed for framing openings, and a wood lintel.

heartwood: The wood extending from the pith to the sapwood.

jamb: The side and head lining of a door, window or other opening.

joint: Space between two adjacent pieces joined and held together by some form of fastener.

joint cement or compound: Can be a powder, mixed with water, used to fill joints and nail dimples on wallboard. Today, most cement is ready mixed.

joist: One of a series of parallel beams, often 2 inches thick (nominally), used to support floor and ceiling loads. Joists are supported in turn by larger beams, girders or bearing walls.

kiln-dried lumber: Lumber that has been oven-dried to a moisture content of 12 percent or less. Today, there is some contention that air-dried lumber remains stronger through its lifetime.

knot: The part of a branch or limb that appears on the edge or face of a piece of lumber.

lath: Material of wood, metal, or insulating board fastened to a building frame, originally used as a plaster base, but mostly used now as a form of shim stock.

ledger strip or board: A piece of lumber nailed along the bottom side of a girder, on which joists rest.

lintel: A horizontal structural member providing support over an opening.

masonry: Stone, brick concrete, concrete block and similar building components bonded together with mortar to form a mass.

millwork: Building materials made of finished wood and made in millwork plants. The category includes inside and outside doors, window and door frames, moldings, interior trim and porch trim.

miter joint: Joining of two members at an angle that bisects the joining angle. As an example, door trim is usually mitered at 45 degrees to bisect the 90 degree angle of the trim.

moisture content of wood: The weight of the water contained in the wood, usually expressed as a percentage of the overall weight of the wood.

mortise: A slot cut into a board or plank to receive the tenon of another board or plank to form a joint.

nonbearing wall: A wall supporting no load other than its own weight.

on center: Measurement of spacing for studs, rafters, joists, posts and other members from the center of one member to the center of the next.

panel: A thin, flat piece of wood or plywood, which may be framed with other material.

partition: A wall subdividing a space within a building.

penny: Originally, the price per hundred nails, but now the term serves as a designation of length. Abbreviated d.

pier: A column of masonry, usually rectangular, used to support other members.

pitch: The slope of a roof, or the ratio of the total rise to the width of the structure. An 8-foot rise on a 24-foot wide house is a ⅓ pitch roof. Roof slope is expressed in inches of rise per foot of run, such as 4 in 12, 7 in 12, etc.

plate: The *sill plate* is a horizontal member attached to a masonry wall. The *sole plate* is the bottom horizontal member of a frame wall. The *top plate* is the top horizontal member of a frame wall and may support beams, joists, rafters, etc.

plumb: Vertical.

ply: Term showing the number of layers of materials such as roofing felt, veneers in plywood, or layers of any built-up material.

plywood: Wood made of three or more layers of veneer joined with glue. The grain on adjoining plies is usually at right angles. Most of the time an odd number of plies is used to provide balanced construction.

preservative: A substance that will prevent action of wood fungi, termites and other destructive agents. It is best applied under pressure.

rafter: One of a series of structural members of a roof. Rafters for a flat roof may also be referred to as roof joists. A valley rafter forms an intersection at an internal roof angle. Normally a valley rafter will be made of doubled members.

rail: Upper and lower members of a balustrade, extending from one vertical support to another.

reinforcing rods: Metal rods or mesh placed in concrete slabs to increase strength.

ridge: The horizontal line formed by the junction of two sloping roof surfaces.

ridge board: A board placed on edge at the ridge of the roof, onto which the upper ends of the rafters are fastened.

rise: The vertical height of a step or flight of stairs.

riser: Vertical board closing the space between treads of stairs.

roll roofing: Roofing material, of fiber saturated with asphalt, that comes in 36-inch wide rolls with 108 square feet of material per roll. Weight ranges from 45 to 90 pounds per roll.

roof sheathing: Boards or sheet material fastened to the rafters on which shingles or other roofing material is laid.

run: The horizontal distance covered by a flight of stairs.

screed: A strip of wood used to give the first smoothing to poured concrete.

seasoning: Removal of moisture from green wood to improve ease of working with the wood, as well as serviceability.

sheathing: Structural covering of wood boards, plywood or other material used over the studs or rafters of a building.

shingles: Roof covering of asphalt, wood, tile or slate cut to stock lengths, widths and thicknesses.

siding: Finish covering of an exterior wall of a frame building made of wood, metal, plastic, etc.

sill: The lowest member of the frame in a structure. The sill rests on the foundation and in platform framing supports the floor joists. It is also the lower member of an opening such as a door or window.

soffit: The covering for the underside of a cornice.

span: Distance between structural supports, such as walls, columns and piers.

square: One hundred square feet. A unit of measure used for roofing and siding material.

stile: A set of steps over a fence.

stringer: The support on which the stair treads rest.

stud: One of a series of slender wood vertical structural members used as supporting elements in walls and partitions.

termites: Insects that superfically resemble ants and are wood eaters.

termite shield: A shield of aluminum, usually, on a foundation wall to prevent passage of termites.

toenailing: Driving a nail at a slant to the initial surface.

tread: The horizontal board on a stairway on which the foot is placed.

trim: Finish materials such as moldings.

truss: A frame or jointed structure designed to act as a beam over a long, open span while each member is subjected only to longitudinal stress.

underlayment: Material placed under finish coverings such as flooring, or shingles, to provide a smooth finish.

valley: An ininternal angle formed by the junction of two sloping sides of a roof.

Index

Index

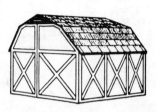